BLOCKCHAIN
& THE SMART
CITIES

Lessons From Singapore,
The World's Smartest Digital Nation

I0479832

SRINIVAS
MAHANKALI

INDIA · SINGAPORE · MALAYSIA

Notion Press

No.8, 3rd Cross Street,
CIT Colony, Mylapore,
Chennai, Tamil Nadu – 600004

First Published by Notion Press 2020
Copyright © Srinivas Mahankali 2020
All Rights Reserved.

ISBN 978-1-64951-766-1

Dedication

I wish to dedicate this book to the numerous sufferers and victims of the horrible Corona Virus Pandemic, nCovid19. I have been inspired by the tremendous success of Singapore in controlling the death rates resulting from this virus to under 0.1% as against 5% across the world. I also dedicate this book to the Singapore Government which has given utmost secure feeling to its citizens while putting the best foot forward in offering excellent medical facilities to its citizens, non-citizens, and the migrant workers alike.

This is because of its excellence in implementing Digital Technologies for the good of the citizens. The best Digital D9 Nation like Estonia and the World's Smartest City State like Singapore stand as the Lighthouses to the world in making Technology work for Good.

I wish the governments across the world will learn from Singapore, one of the most densely populated countries in the world and Estonia, the smallest but finest Digital Nation that have protected their citizens in the best way from day one.

My study about Singapore's best practices has resulted in this book taking shape and I wish to record the best aspects of Singapore's strategic & holistic approach to governance that can serve as a guide to others.

Contents

Author's Information

Srinivas Mahankali,

(https://www.linkedin.com/in/srini-nisg)

Principal Consultant (Blockchain), National Institute for Smart Government

B.Tech. (IIT Chennai 1984-88), MBA (IIM Bangalore 1988-90)

An alumnus of IIT Madras and IIM Bangalore, Srinivas Mahankali is presently a Principal Consultant at NISG's Blockchain Centre of Excellence in a Government & NASSCOM owned organisation, NISG (National Institute for Smart Government), where he plays a key role in promotion, and implementation of Blockchain projects across the Public Sector and Government. He is certified in Lean Six Sigma, NCFM Level 2, Capital Markets and R3 Corda.

Prior to this, he was leading the Blockchain Center of Excellence at ULTS (ULCCS Group, Calicut, Kerala), leading a pool of Blockchain experts to execute solutions for Enterprise & Educational applications.

He also led Strategy & IT for 3 years at Apollo LogiSolutions (ALS), India's leading integrated logistics services company. Here, he formulated & implemented the ABEX (Apollo Business Excellence) program, and executed a comprehensive overhaul of IT hardware, networking and ERP management at ALS. During his tenure as CIO, ALS won the awards, 'The Best Integrated Logistics Services Company' & Best Logistics CEO in India' in 2017.

While he has authored and co-authored two prominent books 'Blockchain–The Untold Story' and 'AI & ML Powered Agents of Automation and Successful Organisations in Action,' respectively, his book 'Blockchain–The Untold Story' is the first ever book to be translated from English into Chinese by Artificial Engineering Bots.

His latest books are Blockchain for Non-IT Professionals and Corona Wars.

A lot of the photos of Singapore landscapes are from his family's personal collections taken during their various trips to Singapore.

Preface

If there is one Magical Technology that can excite everyone with its amazing potential, utility, and benefits for mankind, galvanizing a whole new generation of digital futuristic professionals, it is 'Blockchain.' The last 5 years have seen a rapid change in the understanding of and approach toward Blockchain, the new technological paradigm that has hit the world.

It is now clear to almost anyone that Blockchain not only offers a solution to the many risks being faced by an increasingly centralized and digital world but also is catalysing refreshing changes in the way we work and respond to events.

By acting as a Trusted Third Party in a programmatic manner, Blockchain is bringing in unforeseen trusted interactions, eliminating non-value adding middlemen. This is leading to increased efficiencies and affordability, lowering costs and a vastly improved environment that offers pure, genuine, and authentic interactions, products, services and information for customer delight. Blockchain offers these possibilities by abstracting several technology paradigms that are hitherto simple in nature, by combining them beautifully.

The enterprises and many Governments across the world are finding new use cases of Blockchain every day and it is only a matter of a few more years for us to be totally used to patronizing any service or product that is able to demonstrate its authenticity and purity, an essential feature of a Blockchain solution.

In this book we look at some of the most digitally advanced nations and the status of adoption of Blockchain Technology.

However, stepping more into the world of Smart cities and Digital Nations, we shall see how these countries are leveraging other technologies for the benefit of their citizens.

As a specific use case, we look deep into the world of the Smartest Nation on the planet as judged by the IMD survey, 2019 with a view to offering an integrated road map for the rest of other countries in the world.

Singapore rose from almost nothing 50 years back to be one of the greatest city nations on earth. When I first visited Singapore for a family outing in 1991, Singapore was indeed the dream city for tourists with added attraction for Indians of being a shopping destination for high technology goods, thanks to the tax differential. During my recent visit in January 2020, I was amazed to see the way Singapore stayed relevant and ahead while continuing to be a magnet for talent & tourists by leveraging technology to the hilt. The confidence of the residents in their Government, reflects the way the Governments in the past 55 years have kept the interest, well-being & happiness of the citizens as a prime objective thus winning their hearts.

As I was in awe of everything Singaporean after my trip, I was wondering about how Singapore marched on and on all these years. I also happened to see an amazing documentary on Singapore by National Geographic. One can see it on Youtube with the caption, City of the Future: Singapore – Full Episode | National Geographic, at https://www.youtube.com/watch?v=xi6r3hZe5Tg.

Excellent quality of life, dignity of labour, respect for every person, confidence in the government of every citizen, the warmth in every citizen you meet and greet was palpable, The cleanliness all around, the amazing use of advanced technologies, the vibrant atmosphere and the tourist attractiveness is indeed impressive. The impact of the Government that was invisible but, thoughtful & amazingly effective in every walk of one's life and in charting the course of the ever evolving city that asks 'What next' & faces the challenges by 'Never giving up' seemed significant!

I have been following Singapore & its Government ever since and it is indeed amazing to see the way, one of the most densely populated cities on earth, kept its head cool and won the heart of its people by its calibrated approach to the nCovid crisis through a Multi-Pronged approach. In Singapore, the death rate of one of the most horrible Pandemics we have seen in our times, is by far the lowest in the World at under 0.1% against the global average of over 5%!

Throughout the crisis, no one in the city ever got worked up or lost nerve and became impatient. The amazing use of technology to track & trace the contacts and isolate potential infected persons, the care given to the quarantined and the patients, the support given to the companies and their employees & the way they took care of the migrant workers reflect all the supreme inherent qualities of this great nation, that never rests on its laurels.

Singapore is ranked the world's smartest city as per IMD Smart City Index, an index that measures the way the technology is used to improve the quality of people's lives while reducing the shortcoming of urbanization. In this work, I wish to present some of the amazing facets of Singapore that makes it the most lovable destination.

Singapore, piloted different approaches to implement & accelerate its Smart Nation Vision in its Punggol district, provide a platform for a nationwide roll out with minimal risks and offer better & new ways of living and working, while creating new jobs of the future.

I will dwell upon several facets of Singapore with reference to the emerging technologies and look at how Singapore is leveraging these to serve its citizens to be healthy, happy & smart. There are lessons from these for every other government and country as they grapple with the rapid urbanisation and the need to be self-sufficient & sustainable.

I leverage upon all my experience of over 30 years of working a majority part of it being in CXO positions in Marketing and Technology organisations, that offers me an unique perspective of understanding comprehensively the different aspects and angles that are involved in making Singapore, what it is today.

This is a story of a multi-pronged strategy and a determined effort encompassing all aspects of a society's functioning like business, economics, technology & above all a humane & empathetic approach for a sustained excellence and a never say die spirit.

Acknowledgements

I am extremely thankful to the community of emerging technology professionals with whom I interact on a day to day basis during the past 3 years, working with leading companies and on great projects.

I am indebted to my teachers, senior colleagues and co-authors who have encouraged me to write a series of books on emerging technologies like Blockchain – The Untold Story, AI & ML – Agents of Automation, Secure Chains, Blockchain for Non IT Professionals and a STEM Fiction novel, Corona Wars.

I would like to sincerely thank all those who inspired me and supported me in pursuing my journey in Blockchain. I sincerely thank Ms. Debajani Mohanty, leading Blockchain Author and Practitioner and an ex-Singaporean, who has helped me with her expert advice.

Thanks to Sri JRK Rao IAS, CEO, NISG, and Ms. Debjani Ghosh, President NASSCOM, for continuously and passionately pushing us for emerging technology adoption.

Mr. TN Hari, Head-HR, Big Basket, Ms. Shradha Sharma, Founder, YourStory Media, Mr. Manish Jain, CEO, BPB, Dr. Raghavendra Prasad, CEO, Astra Quark, Raveendran Kasturi, CEO, ULCCS, Mr. Ajit Chauhan, Chairman, Amity Future Academy, my senior colleagues at NISG, MV Ramana Sir and others and Blockchain leaders, entrepreneurs and gurus like Sopnendu Mohanty, Fintech Leader at MAS, Singapore, inspire me to contribute my best at my job and to society.

Above all, I wish to thank my family members with special emphasis on my wife Anuradha and Father-in-law, Sri V.S.R. Murthy, my daughter Deepika Mahankali (Microsoft – Singapore), my son Dr. Sai Prateek and my parents Mr. M.S. Sarma & M. Annapurna who acted as a sounding board to me to crystallize some of my thoughts & also egged me on. They also encouraged me to keep my spirits high through the depressing times of nCovid 19, which proved very difficult.

Topics Covered in the Book

CHAPTER 1: BLOCKCHAIN TECHNOLOGY

1. What is Bitcoin and how Does Bitcoin Work?

2. What is Blockchain? What are the key benefits of Blockchain?

3. What are the different components of Blockchain?

4. Explain decentralisation and distribution in the context of Blockchain application.

5. What are different types of Blockchain?

6. How does Blockchain change the game for the Digital Era Participants?

7. Explain the different cryptographic and programmatic elements of Blockchain

8. Explain the differences between Traditional databases, Distributed Ledgers and Blockchain.

9. How do you define a Blockchain ledger?

10. Describe Blockchain's value proposition across different industries with specific use cases in Digital Identity, Healthcare, Supply chain, Smart city and Government applications.

11. What are the differences between Centralised and decentralised Asset coins?

12. Summarise the different dimensions of Blockchain.

13. What are Smart contracts?

14. What is Ethereum and how does it work?

15. Which are the top 10 countries exploring/leveraging Blockchain to their countries and for what are the key applications & services explored by them?

16. Explain the transaction flow in Enterprise Blockchain applications like Hyperledger Fabric and R3 Corda.

17. How does Blockchain facilitate trusted transactions for secure & scalable automation?

18. What are the key enterprise application areas of Blockchain?

19. Explain tokenisation with examples

20. Explain Central Bank Digital Currency and its benefits

21. Explain the concept of Verifiable credentials on a Blockchain.

22. How do you decide on applicability of Blockchain in an IT problem scenario?

23. What are the top 10 leading Blockchain consortiums in the world and what are they working on?

24. What are the popular use cases of Blockchain in India?

25. How is land pooling scheme implemented on Blockchain?

CHAPTER 2: INTERNET OF THINGS – A KEY BUILDING BLOCK OF SMART CITIES

1. What is IoT and how does it result in machine generated data?

2. What are the different sources of Machine generated data.?

3. What are the different stages of evolution of IoT with time?

4. What are the different components of an IoT system?

5. What are the main layers of IoT platform architecture?

6. What are the important Communication protocols used in IoT system?

7. Describe IoT for individual and Consumer use cases

8. How does IoT power Smart Cities?

9. Explain the relation of other Data technologies with IoT

10. Explain the integration of IoT sensors with enterprise data architecture through a schematic.

CHAPTER 3: SMART CITIES – THE FUTURE OF URBANISATION

1. What is Sustainable Urbanisation?

2. Define Smart city?

3. What are the key steps of the approach taken by the Governments for attaining their Smart city mission?

4. What is the most important component of Smart city and what are its benefits?

5. What are the different layers of ICCC?

6. What are the important activities of Integrated Command and Control centres?

7. Describe The various Smart services offered by the Smart city Governments.

8. What are the different pillars of Smart City infrastructure?

9. Describe the different components of Smart City Business excellence model.

10. Describe 10 Model Smart Cities across the world

CHAPTER 4: HOW BLOCKCHAIN CAN SECURE IOT POWERED SMART CITIES

1. Explain some of the major security breaches across the world.

2. Describe applications where Blockchain is being leveraged to protect the IOT devices?

3. How can we secure robots through Blockchain?

4. How Blockchain can be helpful and used for secured access and management of automobiles?

5. How DRONES can be controlled through Blockchain for security & auditing

6. What are the benefits of Blockchain for Smart cities?

7. Describe Blockchain Real Life Use cases in Smart cities

8. Describe Land Titling using Blockchain in Smart cities

9. Describe Blockchain's role in achieving Sustainable Development Goals (SDG)

10. What are the goals of the Digital Government Strategies as per OECD?

CHAPTER 5: DIGITAL TRANSFORMATION AND DIGITAL GOVERNMENT STRATEGIES

1. What is Digital Transformation

2. What are Digital Twins?

3. What are Gemini Principals and what is their expected outcome?

4. What is Digital Welfare?

5. What are the three major theories of Digital Transformation?

6. What are the characteristics of Digitally advanced organisations?

7. Describe the Digital Transformation Business Excellence Model.

8. What are the 4 main pillars of Digital Transformation and how they combine for facilitating Secured automation?

9. What are DIGITAL NATIONS?

10. What are the common areas of cooperation of Digital Nations?

11. What are the key objectives of Digital Nations?

12. Describe how the top 5 Digital Nations are implementing Digital Transformation strategies?

CHAPTER 6:

1. Explain the visionary architecture of Singapore for Under & over ground infra.

2. Explain is Singapore planning Water Sustainability & managing Pollution Check.

3. Explain the 5T strategy of Singapore to be a global magnet

4. Explain how Singapore encourages Innovation through Sandboxes

5. Explain how Singapore leverages IoT and AI for healthy food for its citizens

6. Explain Hydroponics and its role in Food sustainability of Singapore

7. Explain the role of Smart Nation activities of Singapore and its activities

8. Explain how Technology is used in Singapore for educating youngest to the oldest

9. Explain the various initiatives of Singapore in the areas of IoT, AI & Blockchain

10. Explain the various activities that make Singapore the world's Smartest

CHAPTER 1

Blockchain Technology

Introduction

Blockchain Technology has proved its utility beyond its original discovered use case as a unit of decentralised, distributed and Permission less crypto currency and is not being widely looked upon as a foundational technology that is disrupting a number of industries with a variety of use cases for government and enterprises. With its promise of acting as Trusted Third Party governed by automated programs driven by mathematical algorithms, Blockchain is promising to eliminate expensive non-value adding middlemen who add to a number of leakages of money and a variety of other resources that add significantly to costs. As an inter-enterprise collaborative platform, it is promising to take economies of scale to a totally different scale, benefiting the entire participating eco-system.

The technology has the potential to significantly benefit the humanity by dramatically lowering costs and improving trust in transactions through built in transparency aided by almost immutable & tamper-evident transactions.

In this chapter, we shall look at Blockchain Technology fundamentals and a variety of prominent use cases across different domain, in different countries and also look at a number of Consortiums implementing Blockchain solutions

Origin & Evolution of Blockchain

How it all started

Bitcoin protocol that was launched on January 3rd, 2009, the first known application of the Blockchain technology paradigm, reliably provided a solution for achieving such a consensus in distributed systems that create and transact value over the internet without fear of 'Double-spending.'

This problem was formulated into a story called 'Byzantine General's Problem' where a group of nine generals decided to attack a fort they were surrounding, subject to the majority's decision despite being handicapped by improper communication facilities. A 25-year wait after the problem's formulation, Bitcoin successfully demonstrated a solution for the computer systems to achieve Byzantine tolerance even in face of a sizable number of adversaries and adverse conditions.

There are different types of consensus mechanisms like POW (Proof-of-Work), POS (Proof of Stake), DPOS (Delegated Proof of Stake), PoET (Proof of Elapsed time), PBFT (Proof of Byzantine Tolerance), RBFT, RAFT, N2N and many more. A detailed discussion on these various consensus mechanisms is out of the scope of this manuscript and several white papers are available for understanding and evaluating the same.

Bitcoin, the first implementation of the Blockchain paradigm. Blockchain technology was demonstrated successfully through its first use case, 'Bitcoin.' Bitcoin Blockchain is a living example to show that this often doubted and misunderstood technology is a new paradigm that has come to stay with us for a very long term. Bitcoin is the first implementation of Blockchain technology consisting of six primary elements:

a. An updated Distributed ledger replicated across all the peers undertaking transactions through the platform, consisting of the updated status of Unspent Outputs (UTXO) in chronological order.

b. A network of nodes undertaking to verify and propagate the transactions generated by the participants.

c. A group of miners dispersed across the world to mine the transactions to ensure the authenticity of the same, maintaining the integrity of the Blockchain for all times to come, using an automated execution of the protocol defined by the consensus algorithm called 'Proof-of Work.' 'Proof-of-Work' represents the amount of work that the miners undertake by utilizing their computing power and electricity spent, to be eligible for block rewards in the form of newly mined coins as per a predefined formula.

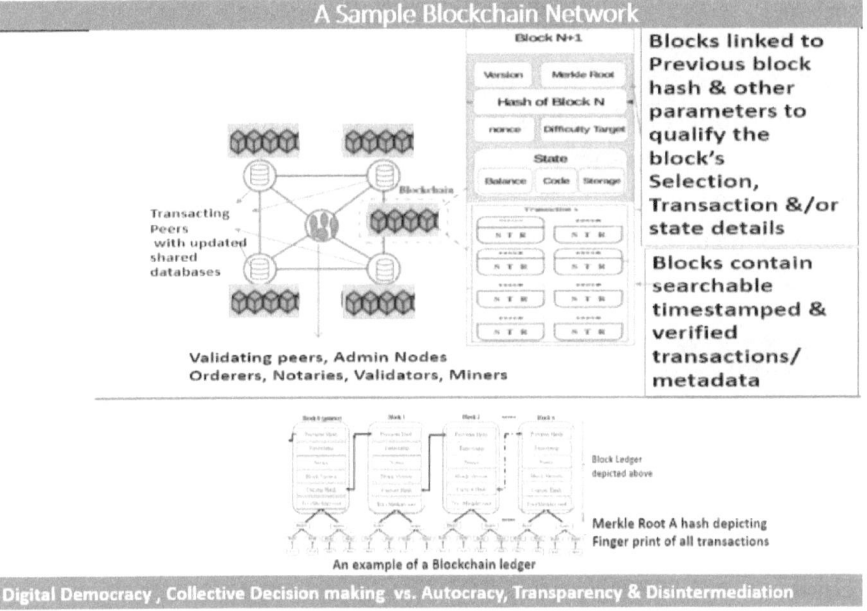

Fig 1.1 Sample Blockchain Network

d. Blockchain wallets used by the participants to initiate transactions and store the value in the form of UTXOs or unspent transaction outputs measured in the number of Bitcoins.

e. The value that is exchanged across the platform, namely the 'Bitcoin' or its fraction, which is treated as a cryptocurrency with all the properties that we associate with the fiat currency in the real world, except the unitized physical representation and regulatory approvals.

f. Exchanges that facilitate buying and selling of cryptocurrencies and derived products known as tokens among themselves using wallets and conversion of the same into fiat currencies in a dynamic manner.

Bitcoin has proved that billions of dollars' worth of value can be exchanged across the world from one person to another unknown person, without the need of a trusted central party, a bank or Government in this case. As on 2nd April 2020, over 18 million Bitcoins with an approximate total value of over 120 billion US Dollar at a unit price of over 6500 US Dollars are in circulation. The success of Bitcoin led to the launch of several variations of alternate Blockchains for a variety of purposes. The majority of them are cryptocurrencies with different properties in terms of privacy, speed of execution, consensus mechanism for transaction validation, the most prominent variation was proposed in the form of the Ethereum Blockchain platform by Vitalik Buterin and his team at Ethereum foundation which we shall discuss a little later.

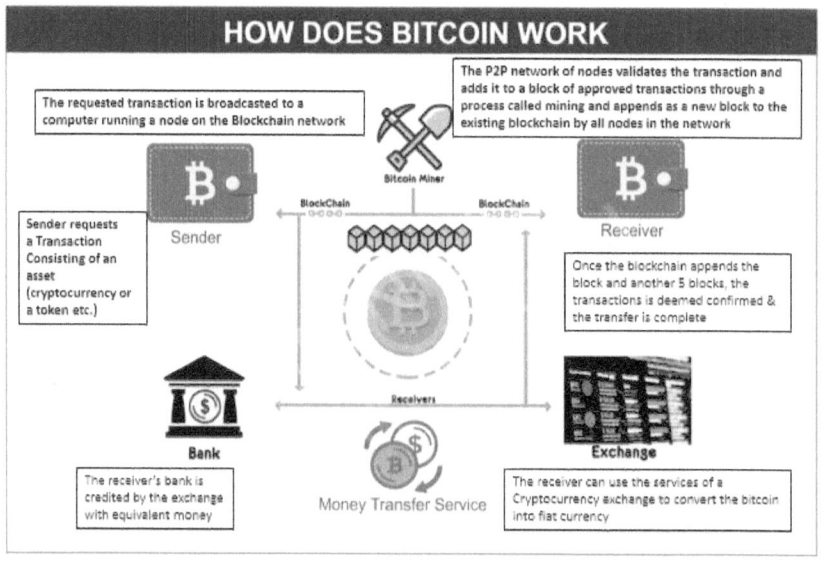

Fig 1.2 How Does Bitcoin Work

Over the years, the cost of infrastructure in the form of storage space and processing power required for IT applications has come down substantially. The availability of high-quality Cloud service providers has reduced the need for investments in high cost on-premise infrastructure. Approaches like 'Open-source technologies,' decentralized methodologies and 'Pay-as-you-go-for-services-consumed' are combining to facilitate the employment of cutting-edge technology powered infrastructure to find new solutions to our problems, rather cheaply. Messaging Protocols, Event-driven communication and record updation, API (Application Programming Interfaces) are facilitating collaboration between applications across multiple on-Premise and Cloud-based applications acting together seamlessly. IBM, Microsoft, Oracle, Amazon and many leading organizations are offering high-end secure IT applications including Blockchain as a service that can facilitate the large-scale implementation of automation enabling technologies in a convenient and cost-effective manner.

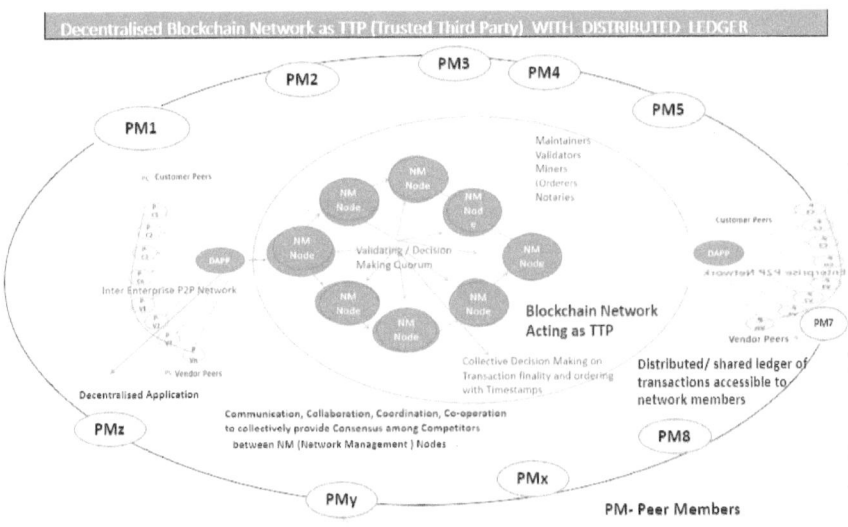

Fig 1.3 Decentralized Blockchain Network as TTP (Trusted Third Party)

Resilient Data Structures of Blockchain

We have seen that in the traditional approach, the participants in a typical business scenario pretty much operate in silos and all the parties are connected to the centralized big market place or the dominant player who connects the buyers and sellers or provides the services to the clients globally.

Instead, Blockchain presents an inter-enterprise scenario that offers a 'Single Source of Truth' where all the peers are connected to every other peer with a possibility to conduct peer-to-peer transactions as per business logic codified in the form of Smart Contracts. Even the dominant player, though while being the facilitator could still be a player whose returns depend on the quantity and quality of the business dealings happening on the network.

The Single Point of Failure has always been the bane of most of the centralized organizations which maintain their databases under a single command, control and administration. This is the weakness most often exploited by the Ransomware virus creators who were behind some of the most lethal attacks on global organizations by unleashing the WannaCry virus.

While distribution and shared database also help in non-repudiation by the parties undertaking transaction, the ability to reconstruct the database from other members of the network eliminates the risk of the SPOF from this very route, thus blunting the weapons of the cybercriminals. This minimizes the risk by tilting the RRR (Risk-Reward-Ratio) away from the investors of these crooked instruments.

Thus, Blockchain is seen as the vehicle for safe and secure automation at scale.

Ethereum which was launched after Bitcoin with improvisations, allowed businesses to create decentralized versions of real-life applications that we see in the day-to-day world through the implementation of 'Smart Contracts' which are programs created to replicate the business agreements into applications that can be run on Blockchain databases.

A "smart contract" is simply a piece of code that is running on Ethereum. It's called a "contract" because code that runs on Ethereum can control valuable things like ETH (ether – native crypto currency on Ethereum Public network) or other digital assets. Smart

Contracts abstract real-life business agreements into applications on a decentralised network of computers running Ethereum nodes in the context of EVM (Ethereum Virtual Machine). The EVM runs as a local instance on every Ethereum node, but because all instances of the EVM operate on the same initial state and produce the same final state, the system as a whole operates as a single "world computer." EVM is considered Turing complete which means it can solve any reasonable complex computational problem. One technically implements logic in say Python and translate to Solidity the Smart contract programming language to implement sophisticated logic.

In the case of Permissionless Blockchains like Ethereum, the transactions forwarded by clients are validated by a pool of miners who win the opportunity to create a block of valid transactions that is then appended to the Blockchain, eventually updated by all the nodes of the platform. Thus, the approved transaction becomes a part of everyone's ledger thus becoming immutable and tamper resistant.

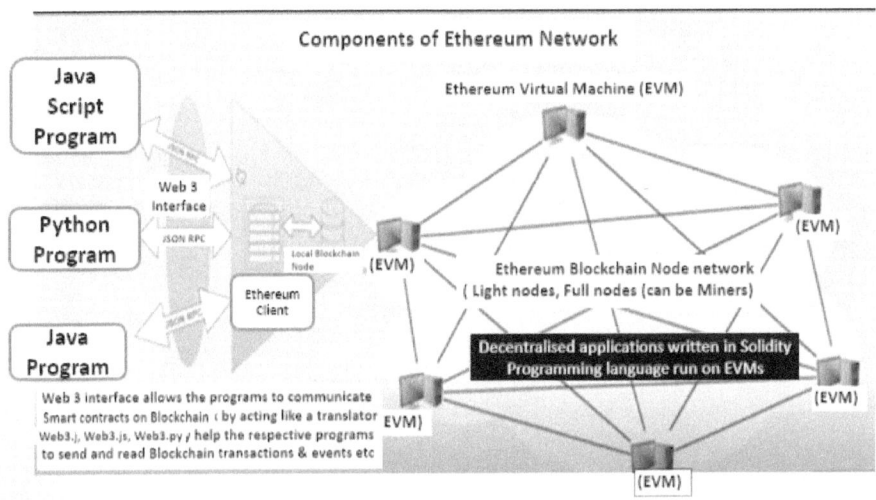

Fig 1.4 Components of Ethereum Permissionless Network

Enterprise Blockchain platforms like Hyperledger Fabric, Quorum, etc., were developed as variations of the Ethereum platform while enterprise applications like Multichain and R3 Corda took inspiration from the architecture and other elements of Bitcoin Blockchain.

Components of Blockchain

Blockchain Combines Encryption, Encoding, Hashing, PKI, Timestamps, DSA and Broadcast for the Internet of Value by bringing Privacy, Permission, Password management within the reach of an individual peer and frees him/her from the dependence on the Trust Anchors who have now grown unduly large leading to a centralized internet

Blockchain converts the traditional internet infrastructure as we know through its TCP/IP protocol from the Internet of Information to the Internet of Value, by acting as a Trusted Third Party to any peer-to-peer interactions. The features of Blockchain that facilitate this are shown in the following figure.

How Blockchain changes the game for the Digital era participants

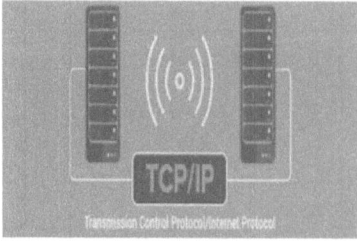

Problems:
- Open communication
- Meant for only information exchange
- No encryption
- Prone to Data breach
- Low trust protocol

Blockchain offers : Encryption, Permission & Access rights, Smart contract capability, Privacy using ZK SNARK, Homomorphic encryptions.

Leads to secure conditional trading of data and value at scale & Trust between unknown entities

- Resulting in **Internet of Value** Versus **Internet of Information**
- Blockchain makes IoT devices Safe and Secure to operate
- Blockchain offers high quality data for AI/ML applications & Increase ROI on Analytics investments

DISINTERMEDIATING TRUST — EMPOWERING INDIVIDUAL USERS — FACILITATING P2P TRANSACTIONS

Fig 1.5 How Blockchain changes the game for the Digital Era Participants

Blockchain combines the cryptographic and programmatic paradigms like encryption, encoding and hashing in a unique manner to achieve amazing benefits offering a new paradigm of trusted disintermediated transactions.

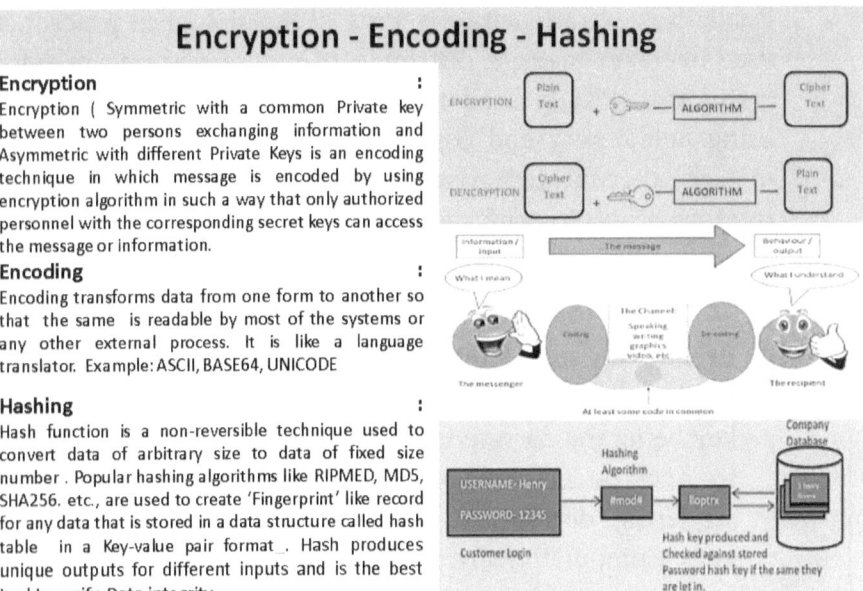

Encryption - Encoding - Hashing

Encryption :
Encryption (Symmetric with a common Private key between two persons exchanging information and Asymmetric with different Private Keys is an encoding technique in which message is encoded by using encryption algorithm in such a way that only authorized personnel with the corresponding secret keys can access the message or information.

Encoding :
Encoding transforms data from one form to another so that the same is readable by most of the systems or any other external process. It is like a language translator. Example: ASCII, BASE64, UNICODE

Hashing :
Hash function is a non-reversible technique used to convert data of arbitrary size to data of fixed size number . Popular hashing algorithms like RIPMED, MD5, SHA256. etc., are used to create 'Fingerprint' like record for any data that is stored in a data structure called hash table in a Key-value pair format . Hash produces unique outputs for different inputs and is the best tool to verify Data integrity

Fig 1.6 Encryption – Encoding – Hashing

i. **Hash function:** A hash function produces a unique fixed length output for a corresponding input of any size (like a thumb impression of a human being), which cannot be replicated. Hash of any information is treated as the unique and indisputable representation of the information. Hashes form the heart of Blockchain as the blocks are represented by the hash of the information and are chained together as a linked list of chronologically mined and validated blocks.

ii. **Merkle root (Root of roots):** While a hash is a unique number derived out of the base number, the Merkle root is derived from hashing pairs of transactions together until only one element is left. Since the hash was unique, a change in any transaction would result in a change in the Merkle root, which would be easily caught.

iii. **Public-Key Infrastructure:** To facilitate secure electronic transmission of information and undertake ultra-safe transactions, Blockchain employs several cryptographic applications. PKI or Public-Key Infrastructure is a set of technological procedures used to create, manage, distribute, use, store, and revoke digital certificates. PKI is used to authenticate participating parties using public keys and corresponding private keys connected to each other through complex algorithmic relations, requiring rigorous proofs to confirm identities for facilitating information exchange. PKI uses X.509 certificates to identify the owners of public keys.

   a. **Private key and Public-Key:** The Private Key and Public-Key pair is (Private key being the secret password and Public key being the corresponding username known to all) used to encrypt information using mathematical algorithms, rendering decryption virtually impossible without these keys. Computationally, it is similar to the factoring of prime numbers, which is a simple, mathematical procedure. However, decomposing the result is difficult without prior knowledge of its factors.

   b. **RSA:** PKI systems normally use RSA algorithms for linking public keys and private keys. RSA (Revest–Shamir–Adleman) is one of the first public-key cryptosystems and is widely used for secure data transmission. In such a cryptosystem, the encryption key is public and it is different from the decryption key which is kept secret (private).

   c. **ECDSA:** Blockchain systems use Elliptical curve cryptography to issue secure Public-key Private key pairs. The messages are encrypted by a digital signature algorithm namely, ECDSA that ensures that only authorized owners of targeted messages can securely decrypt the messages.

iv. **Digital Signatures:** Digital signatures are a unique aspect of Blockchain transactions and provide a layer of security to carry out and validate genuine transactions. A digital signature is a mathematical scheme to present the authenticity of digital messages or documents. A valid digital signature gives the recipient reason to believe that the message was created by a known sender (authenticated by verifying against the public key of the sender), and the sender cannot deny having sent the message (non-repudiation by signing with his/her unique Private key), or that the message was not altered in transit.

v. **X.**509 Certificates: In Permissioned Blockchains like R3 Corda or Hyperledger Fabric, the participating members are provided X.509 certificates by the administration Certificate authority for identification by the network. An **X. 509 certificate** is a digital **certificate** that uses the widely accepted international **X. 509** public key infrastructure (PKI) standard to verify that a public key belongs to the user, computer or service identity contained within the **certificate.**

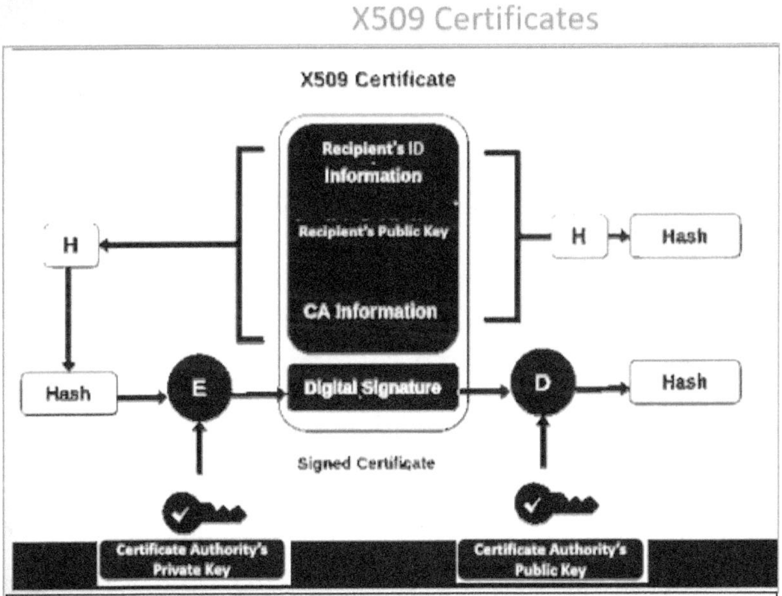

Certificate consists of 2 parts- 1 part contains the information about the Recipients' details and the Certificate authority's information and its hash. Another part is the hash of above information signed by Certificate Authority's private key that can be decrypted by the CA's private key. When the Recipient presents the X.509 certificate to the 3rd party, the verifier decrypts the 2nd part, and matches the hash with the hash of the information in the first part. If it matches, the recipient is verified as also the Certifying authority.

Fig 1.7 X.509 Certificates for certifying identities in permissioned scenarios

vi. Consensus Mechanisms (POW, POS, DPOS, PBFT, etc.): The mechanism by which members come to an agreement about the authenticity of a transaction is referred to as the 'Consensus Mechanism.' Consensus formation ensures the involvement of multiple validators in a systematic and predetermined manner, ensuring decentralization and objectivity of decision-making. It ensures implementation of the key features of the Blockchain platform like increased trust, immutability of the transactions, and maintenance of the integrity of the platform. The consensus mechanism is the soul of the Blockchain platform and has to help members in reaching the right decision all the time. The sanctity of the Blockchain application depends on the strength and reliability of the consensus mechanism. The consensus mechanism followed by Bitcoin and the earlier version of the public Ethereum client is known as 'Proof-of-Work (POW)' where miners or validators compete with each other and burn valuable resources like computing power and enormous amounts of electricity to guess the right Nonce (number used only once) and create a targeted hash to win the race to create a block. Proof-of-Work—followed by Bitcoin Blockchain and some versions of Ethereum Blockchain—consumes a huge amount of resources to arrive at a deterministic consensus. The Ethereum platform will soon shift to a 'Proof of Stake' based consensus, which involves negligible energy consumption.

Some new-generation public platforms use variations of 'POW' - and 'POS'-based consensus algorithms like PoET (Proof of Elapsed Time) and DPOS (Delegated Proof of Stake) to minimize resource utilization and wastage. Enterprise Blockchains use energy-efficient algorithms like 'Proof of Authority' (POA), Practical Byzantine Fault-Tolerant' (PBFT), 'Node to Node' (N2N) and their variations to arrive at a deterministic consensus. Blockchain's Magical Components are described in the following figure:

MAGIC CALLED BLOCKCHAIN

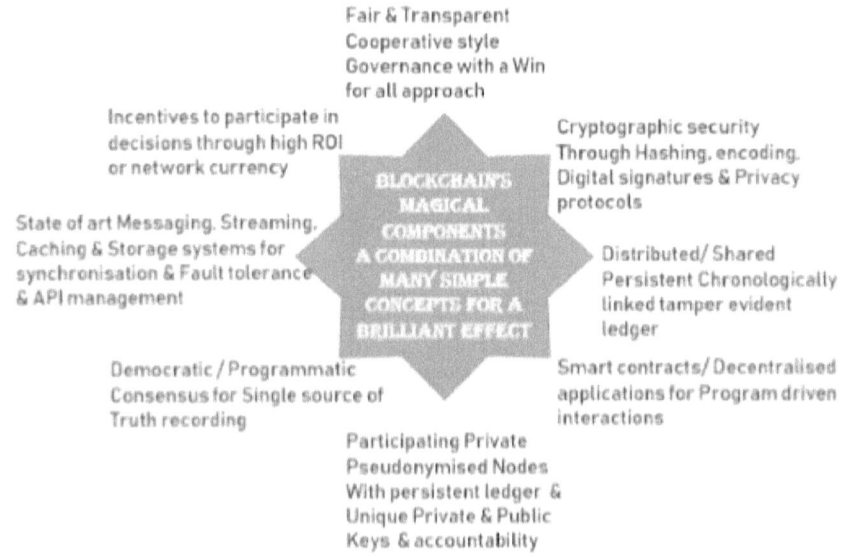

Fig 1.8 Magic Called Blockchain

As seen in the above section the discovery of the Blockchain paradigm has been achieved by an ingenious combination of the various simple tools and techniques that have been in vogue for decades. Let us now define Blockchain with our understanding of the various components, features and benefits offered by this unique technology.

Blockchain can be thus defined in the following manner with all the various components.

Blockchain – An augmented Distributed Ledger Technology

Blockchain is a Distributed Ledger that uses peer to peer consensus within a Decentralized Network to validate transactions and a hashing algorithm to cryptographically link them in a Chronological chain of records.

1. Copies of ledger are shared across computers known as 'Nodes in the network'
2. Computerised record of historical transactions chronologically ordered.
3. Shares resources directly between nodes bypassing third part network with specialized communication protocols.
4. Every transaction must be approved (or rejected) by Consensus mechanisms (Ex: POW, POA, POS, DPOS, PBFT, RBFT, Raft etc.)
5. Hosted by many nodes simultaneously controlled by no single entity. Data is accessible by anyone within the network.
6. Public Permissionless networks like Bitcoin & Ethereum, Private Permissioned networs like Hyperledger Fabric, R3 Corda, Quorum
7. Transactions may include moving currency, updating a standard enterprise records, transferring ownership of an asset etc.
8. Converting transaction data to a fixed length string of numbers and letters that cannot be reverse engineered (ex: SHA 256).
9. The hashing process of a new block includes meta data from the previous block's hash output. The link makes the chain immutable.
10. A full immutable time-ordered history of transactions approved by the network.

Fig 1.9 The different facets of Blockchain Paradigm

Definition and Types of Blockchain

What is Blockchain? Blockchain is an augmented Peer to Peer Distributed Ledger Technology employing advanced cryptography to secure identities of participants in the network undertaking timestamped, immutable transactions with decentralised processing to exchange data & change ownership of assets using cutting edge technology powered applications also known as Smart contracts running inside the system providing Transparency, Security, Tamper resistance, Auditability and enhanced Trust through system acting as Trusted Third Party in Triple entry accounting.

Traditionally we are used to centralised databases for storing data which respect the CRUD methods for manipulating data, namely Create, Update, Update and Delete. We also come across replicated data bases under the command and control of the same IT Admin control.

Blockchain, a programmable database differs from the traditional databases in a number of ways, as captured in the following figure.

Blockchain and its connection with Databases and Distributed Ledger Technology:

Blockchain, Distributed Ledger Technology & Databases

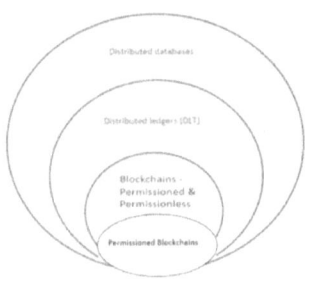

A. Distributed Ledgers exclude Replicated Databases.

B. Distributed Ledgers need not be linked to Chain of Blocks and can be updated by authorised & approved transactions

C. Blockchains are a **Special type of Distributed Ledgers** that
1. Contain Transactions that are agreed as per consensus algorithms
2. Approved transactions are timestamped & cryptographically linked and the records are shared in blocks.
3. **Permissionless** Blockchains are symmetrical with equal data and rights to all
4. **Permissioned** Blockchains contains selectively but mutually shared transaction records and hence nodes can have different cumulative ledgers.
5. **Public** Blockchains have data that is visible to all while

The relationship between blockchain and DLT (Cointelegraph n.d.)

Private Blockchains have data visible as per access controls

Fig 1.10 Blockchain & its connection with Distributed Ledger Technologies and Databases

As seen in the above diagram, Blockchains can be classified as

i. Permissioned – If the membership and the validator pool is restricted and must be approved by an admin authority like in the case of Hyperledger Fabric, R3 Corda, Quorum etc. In a Permissioned Blockchain, ability to conduct transactions or write data is restricted as per access control rights

ii. Permissionless – If the membership and the validator pool is not controlled and accessible with equal opportunity to anyone like in the case of public Bitcoin and Ethereum platforms

In case the data stored on the Blockchain is accessible for viewing to anyone without restrictions, then it is considered a Public Blockchain and if the access is strictly restricted and is kept confidential to a selected group of participants, then it is considered Private Blockchain.

If a Blockchain is set up and implemented by a dominant player who controls the access and validation, it is termed Private Permissioned and in case a group of participants work together then it is termed a Consortium Blockchain or Federated Blockchain.

Thus, many permutations & combinations are possible depending on the ability to read, write or vote on the transactions.

In the case of Governments, we come across Public Permissioned Blockchains which are restricted for writing and maintaining, but the data could be accessed by all the citizens for verification like in the case of certain type of certificates or ownership records.

The important feature of this Blockchain approach is the 'decentralized' approach where the decision regarding the correctness of the transactions is taken without recourse to an individual entity's authority and muscle power. The transactions with due approvals and authorizations representing the real-life scenario are sent to a pool of network managers, who can then collectively follow a designated approach and vote on the transactions to be included in the approved chain of events that influence the records and ledgers permanently.

The decentralized pool of miners is referred differently in different Blockchain systems and serves to increase the uptime of the network manifold while minimizing the risk associated with a centralized approach. While in Permissionless Blockchains we have mining pools or set of validators, on Permissioned Blockchains for enterprise applications, they operate as a set of Orderers (Hyperledger Fabric), Notaries (R3 Corda), Validators (Hyperledger Sawtooth, Indy) and the like.

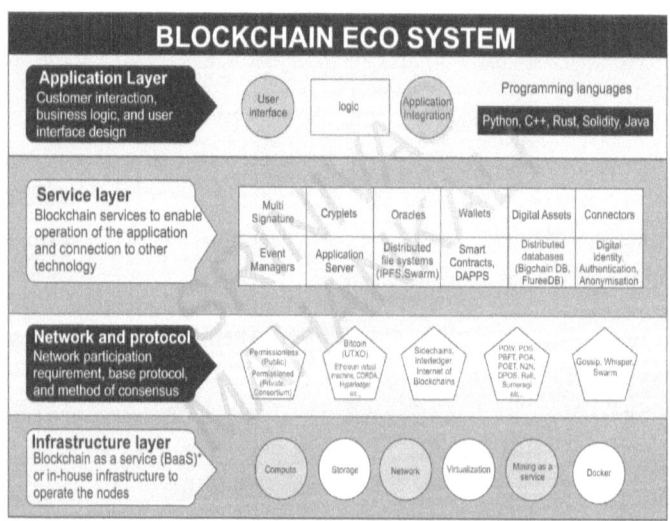

Fig 1.11 An integrated representation of different components of Blockchain eco-system

Though these networks start with single such nodes in the beginning, in multi-enterprise scenarios, they operate in a pool (also with replication for Disaster Recovery purpose) and follow appropriate consensus mechanisms depending on the requirement by the administrators of the network.

The network offers 'Decentralization' at different levels. The network effect of combining multiple transacting parties into one channel creates a technically decentralized system for pushing the transactions through a client and the validation and ordering of these transactions is undertaken in a decentralized manner by the Blockchain infrastructure.

Hyperledger Fabric works on the concept of Channel, which is a private network within the quorum of all the nodes on the system, that share a business logic and are parties to transactions as per an approved smart contract also called a Chain code. The data is shared amongst the participants as per access control and privacy requirements to maintain confidentiality unlike in the case of Ethereum which broadcasts the information on the ledger to all the participants.

Transaction flow in typical enterprise Blockchains like Hyperledger Fabric is described in the following figure:

Hyperledger Fabric- Typical Transaction flow

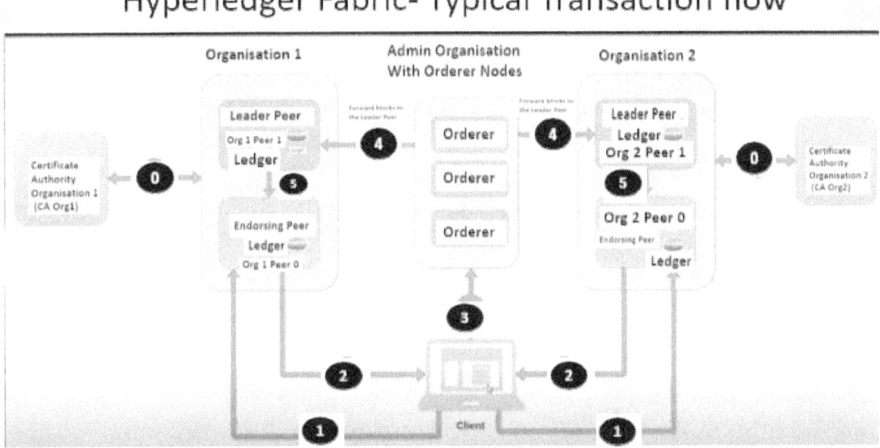

Fig 1.12 A typical transaction flow in a Permissioned Blockchain Hyperledger Fabric Network's Channel

Step 0: The Certificate authorities in the respective organisations provide the cryptographic identities to the respective peers and the same information is disseminated to all the other counterparties involved in transactions with the respective peers.

Step 1: Client sends Transaction Proposal to Endorsing peers (who must approve as per Business logic encoded in the channel's Chaincode).

Step 2: Endorsing peers attest the transaction and send back to Client with their digital signatures

Step 3: Clients sends the fully approved transactions along with the endorsers' signatures to Orderer

Step 4: Orderer verifies the transaction's validity and includes the same in a block along with the time stamp. As per the block interval/block size limit encoded, Orderer creates a block of valid transactions and sends the read/write sets to respective organisation's Leader Peers.

Step 5: Leader Peer distributes the blocks to all the peers in the organisation who are the channel members

Step 6: The validating peers upon receipt of the blocks, update their respective ledgers with valid approved transactions consistent with their current state

Step 7: Transactions not consistent with the current state of the respective member ledgers are nullified but continue to be a part of the Blockchain ledger

Distributed Ledger Platforms like R3 Corda do not work on the concept of Blockchain.

They follow the Triple entry accounting concept, where a Notary node acts as a validator to guarantee transactions between counterparties, as per pre-configured business logic. Notary checks and ensures the validity of the transaction and prevents double spend.

Nodes run flows (mostly) to update the ledger

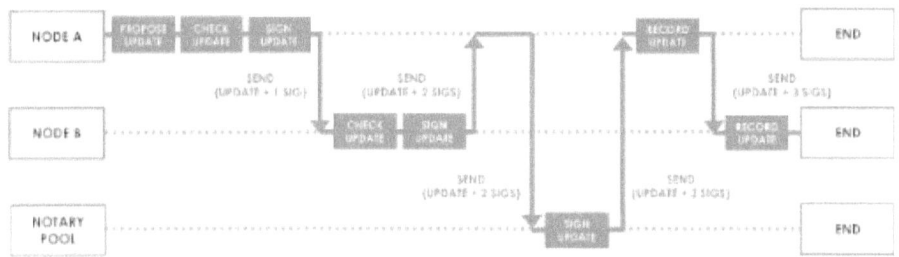

Fig 1.13 R3 Corda Distributed Ledger system operates as per Triple entry accounting system on a transaction to transaction basis

In general, Blockchain databases can be considered SALT databases as per the context may be. In the context of Permissioned Blockchain systems, SALT may be described as Sequenced (timestamped), Agreed (decided in a manner agreeable to the participants as per an approved program), Ledgered (maintained in a database of key-value pairs reflecting the state of ownership of assets and Tamper resistant (almost impossible to change the order of the records committed).

While the existing incumbents involved in running the businesses across the Governments are comfortable with the centralized approaches, the choice of a decentralized approach and Distributed Ledger Technology for the future is seen as a decision that may yield substantial returns, but are fraught with unforeseen risks. As the technology is still in its early stage of adoption, several factors need to be considered by the decision-makers to undertake the decision to migrate to the new paradigm. The following table gives a bird's eye view of the aspects to be considered for evaluating the suitability of a Distributed Ledger Technology-based solution

Do You Need a Blockchain?

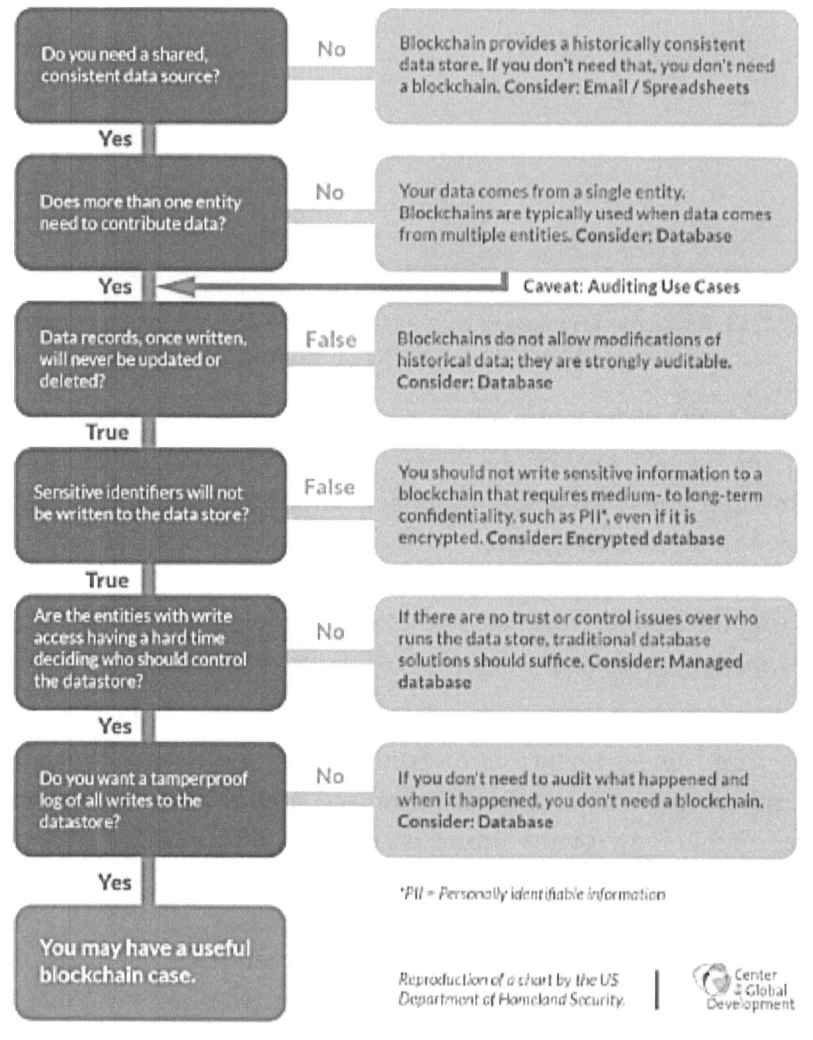

Fig 1.14 Do you need a Blockchain? Decision flow chart, courtesy

Source: https://www.cgdev.org/publication/reassessingexpectations-Blockchain-and-development-costcomplexity

A high-powered committee with the involvement of the top management professionals should consider and analyze the various aspects of the problems to be tackled and evaluate the potential solutions.

Applications of Blockchain Technology

Blockchain's key applications are well be summarized in the following figure:

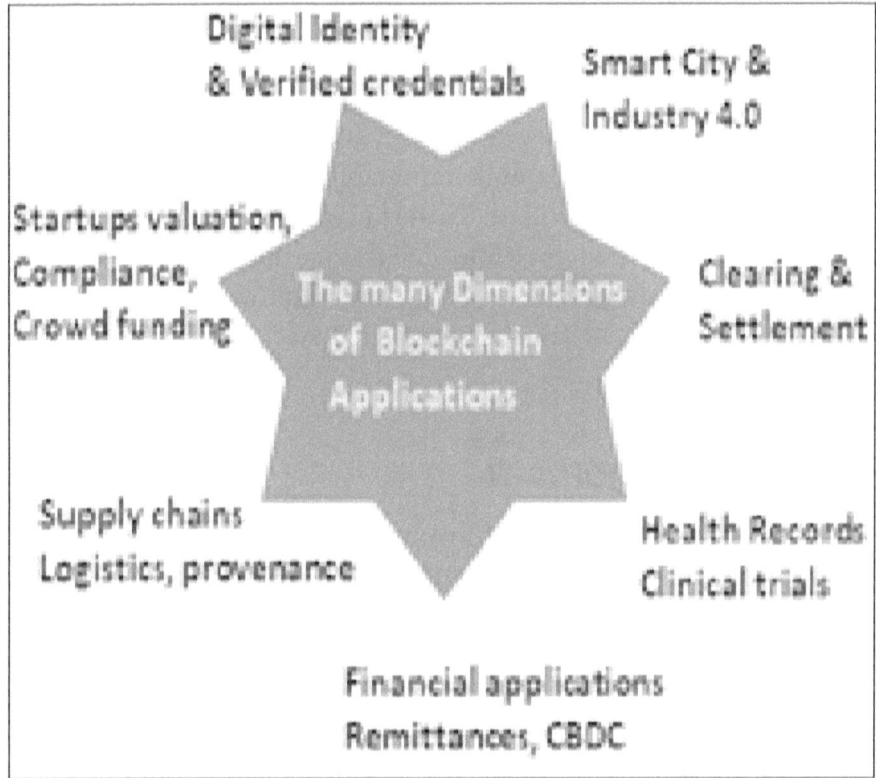

Fig 1.15a Key applications of Blockchain

Blockchain's Value Proposition across Industries and use Cases Government

Problem: Governments offer numerous certificates of identification and authenticity to their citizens. They are the biggest spenders of public money, collector of taxes and distribute subsidies. Governments also undertake large projects on an ongoing basis inside the country and outside the country through the External Affairs Ministry, Government must implement Smart City projects and provide cybersecurity to critical installations. All these areas are fraught with complexity and potential frauds leading to a lot of leakage and loss of money on a huge scale.

Solution offered by Blockchain: By offering fool proof methods for issuing unique digital identities, ensuring the provenance of goods and supplies, providing certificates registered on a Blockchain and ensuring accountability and transparency in project management, procurement, vendor management and supply chains through non-refutable digital signatures and immutably stored data that cannot be manipulated and modified, Governments stand to save a huge portion of their expenses while delighting the citizens. After all citizens demand transparency, efficiency, ease of interactions and lower costs that Blockchain's dis-intermediated trust can offer as a de-facto outcome!

Project Management

Problems: Project Management involves the delivery of expected and planned outcomes through the utilization of defined and budgeted resources comprising of money, manpower, materials and time. This involves interactions between multiple parties both inside and extraneous to organizations and the results are dependent on multiple parties working strictly in consonance to their contracts relating to the implementation. Most often, the lack of synchronization between parties concerned and participants falling short of their deliverables lead to cost and time overruns and humongous losses throwing all the plans awry.

Solution offered by Blockchain: Recording the Contracts and monitoring the project status and adherence to the deliverables in the same manner by all the parties concerned, as per milestones, is best done over a Blockchain platform to ensure compliance. Cryptographic references to project status are stored on the Blockchain and the status report with respect to the deliverables are shared periodically and on critical matters shared over the Distributed ledgers in real-time. The tamper-evident nature of the records and the 'Triple entry-accounting' feature of the Blockchain acting as a Trusted Third Party will ensure that the accountability of all concerned parties is monitored in a fool proof manner for timely action and also to dispense rewards/penalties required to put the projections execution ahead of schedule on all dimensions. Transparent procurement process, automated bid management with details of successful vendors recorded on a Blockchain and monitoring the progress of their deliveries can also trigger delivery versus payments in a trusted manner with the highest accountability of the involved officials, thus eliminating chances of errors and misappropriation as well.

Digital Identity

Problems: Multiple records, Duplication of efforts and processes, Siloed systems and potential for identity fraud and that of stolen credential copies.

Solution offered by Blockchain: Issue and verify once on Blockchain, link multiple identities to a unique Blockchain identity-operated through a single user interface or a digital wallet, eliminate the need for multiple verifications across establishments thus saving a lot of time, effort and documentations which maximizes the trustworthiness of the identity information.

Voting

Problem: Tedious manual paper and printing intensive processes requiring humongous funds and fake/unaccounted identities pose enormous challenges for countries and enterprises undertaking elections for governing bodies and on-board resolutions.

Solution offered by Blockchain By uniquely identifying voters in a fool proof manner and recording their votes by their digital signatures through a verifiable and non-refutable system, Blockchain eliminates fake votes, wrong votes and extensive paperwork eliminating wasteful processes to reduce costs enormously.

Registries and Certificates

Problem: Fake certificates, high cost & time required for issuance and verification plague documentation of events from birth to will execution for asset acquisition and credential accumulation.

Solution offered by Blockchain: Educational Municipal, Police and other credential certificates can be issued and shared securely eliminating fakes and offering benefits for instant audit and reconciliation while establishing clear title.

Benefits and Subsidy Distribution

Problem: Fake claims, excessive middle layers leading to leakages and adding non-value costs drain valuable resources of Government and trusts.

Solution offered by Blockchain: Clear identification of beneficiaries, allotment and monitoring of benefit utilization for every unit issued with minimal intermediary intervention in near real-time allows for high productivity of welfare spends.

Supply Chain

Problem: Procurement: Subjectivity and opaque procurement processes create leakages and mistrust. **Financial Documentation:** Letter of Credit, Suppliers credit and other financial transactions offer a lot of scope for manipulation and mistrust. **Provenance:** Fake goods and wrong claims for premiumness hamper a variety of goods ranging from Pharma, food, imported, exported and specialized products **Retail:** Warranty claims, Loyalty rewards cross multiple vendors are difficult to track and often lead to disputes **Transport conditions:** Un-monitored cold storage transported goods like pharmaceuticals, food, milk and dairy products lead to the consumption of spurious/expired products.

Solution offered by Blockchain: Transparent and Trusted processes offered by immutable, shared ledger of records between verified identities. Digital signatures for non-repudiation and shared ledger for near-real-time communication drastically reduces costs and scope for frauds. Smart contracts triggered to capture the events like a change of ownership and transfer of assets immutably on a shared ledger, help identify the origin of the products along with certifications of the originality of standard adherence, especially valuable in Automotive spares. Blockchain facilitates seamless tracking of warranty claims and allotted rewards until redemption for increased effectiveness and benefit of consumers. By recording the temperature of cold-stored items across the supply chain and tracking them on a Blockchain ledger, the consignment details of spoiled items can be quickly traced. This will minimize the propensity of wilful manipulation.

Health Care

Problem: Fake drugs, Compliance in Clinical record management, health record tracking and settlement of insurance claims are often causes for fraud and manipulation.

Solution offered by Blockchain: Blockchain can offer multiple benefits for solving the various challenges of health care domains like seamless management of EHRs with utmost privacy and security features, transparent compliance tracking in case of clinical records and insurance settlement and Origin-to-chemist tracking of Pharma goods, etc.

Smart City

Problem: Unauthorized access by cybercriminals to leverage net connectivity of the IOT devices for DDOS attacks and illegal actions like crypto-jacking, data leaks, etc. The command and control of autonomous vehicles and drones need to be secured against cybercriminals.

Solution offered by Blockchain: Blockchain offers a protective shield for IOT Gateways, autonomous vehicles, drones, and robots and prevents unauthorized access by criminals and manipulators. This enables secured automation. Blockchain facilitated accurate assessment of renewable energy claims and peer-to-peer energy trading among Prosumers.

Further, Blockchain has extensive uses in a Smart city scenario in securing, Smart healthcare, • Smart transportation, • Smart energy, • Smart government, • Smart tourism, • Smart education and • Smart environmental protection

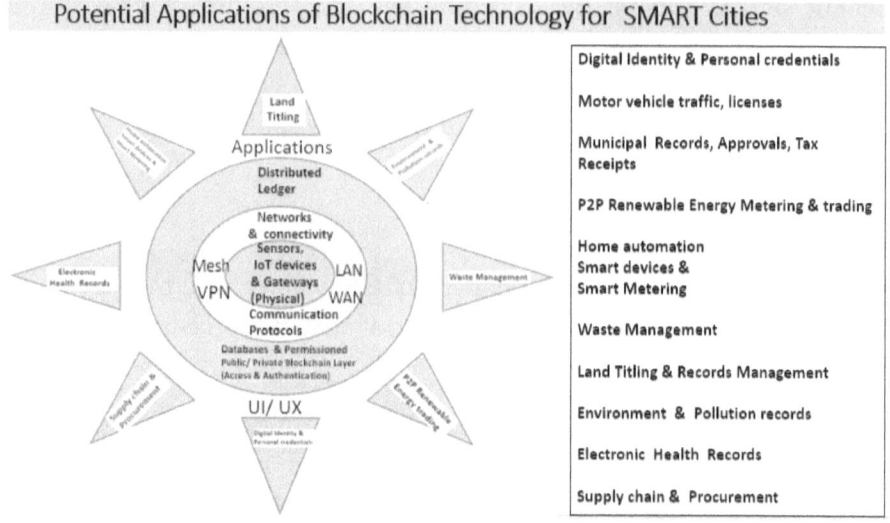

Fig 1.15b Blockchain's applications in Smart Cities

Cybersecurity

Problems: Single points of failure of centralized management offer valuable targets for cybercriminals. Increasingly digitization and billions of internet connections managed by centralized systems run the risk of derailment and ransom attacks. WannaCry, one such virus infected 230,000 computers in over 150 countries, using 20 different languages took $300 US Dollars per computer to decrypt and release the data.

Solution offered by Blockchain: By sharing distributing data across multiple ledgers, authenticating identities, encrypting transaction information, Blockchain offers a de-risking mechanism for data-intensive applications and blunts designs of Ransomware criminals who fraudulently sneak into corporate systems, encrypt the data and demand ransom to decrypt the same.

Eliminating Fake Certificates & Identities

The utility of Blockchain in eliminating fakes through trusted document management and ensuring source to destination ownership tracking can be succinctly summarized in the following lifecycle activities that could be authentically stored on a Blockchain:

- **Cradle to Grave/Womb to Tomb** – All certificates in one's life from birth certificates, vaccination records, Health/Property and Academic, Non-academic and Identity records, Will recording and execution, etc., need impeccable tracking that Blockchain provides.

- **Vivad to Viswas (Dispute to Trust)** – Any agreements and compliance issues can be easily reconciled.

- **Farm to Fork/Catch to Consumption** – Safe and compassionate handling of animals and amphibians meant for consumption can be tracked through the supply chain.

- **Procure to Pay** – Complete transparency in the Procurement process by recording activities in every stage. Procurement is the biggest source of subjective behaviour that can be made transparent.

- **Pay to Cash** – Manpower and work outsourcing organizations can minimize Pay-Bill cycle leakages by instant settlements and eliminating the need for reconciliation.

- **Admission to Retirement** – Academic and non-academic certificates and transcripts can be stored and shared privately without any fear of fake certificates and time loss.

- **Segregation of Duties:** In issues of Project management or execution of shared responsibilities in organizations, IT projects and new product development, there is a need for responsible and automated tracking of discharge of one's duties.

Digital signatures and non-repudiation help in achieving instant confirmations and recognition of good and productive behaviour.

- **Start-up valuation and compliance tracking:** Most of the small companies suffer from the inability to capture value contributions and tracking from the promoters and investors. Blockchain enables perfect, real-time valuation, promoter shares' tracking and support in compliance management for the Start-up founders from the idea stage itself.

- **Sanction to Settlement:** Many activities in Government and enterprise domains need approvals and endorsement. Blockchain can track the documentation and attestations from approval to settlement in an impeccable manner. House designs, Police approvals for public meetings, large project budgets are some of the many such activities that can benefit from the Blockchain approach. The following solution depicts a typical document management solution by leveraging Blockchain technology to eliminate fake certificates and facilitate trusted sharing of information guaranteed by Blockchain while protecting from malware attacks and any form of unauthorized tampering.

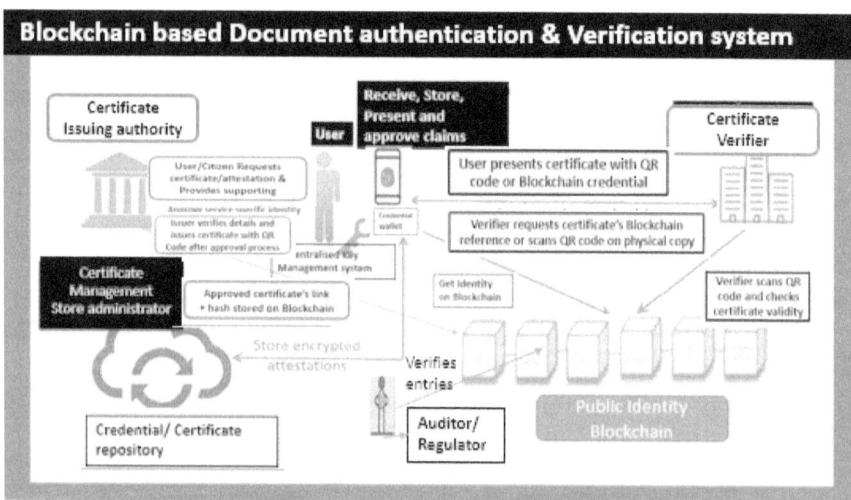

Fig 1.16 Blockchain based instant Document authentication and Verification system is depicted in the diagram

Loyalty, Games and Sweepstakes and many more applications that depend on Trust are lifelong relationships between hitherto unknown parties cutting across domains, will find Blockchain an interesting platform to adopt and provide value to the peers on either side. Blockchain can also help in a variety of Smart City applications to save lives and improve quality of life.

Digital Representation of Assets and Rights on a Blockchain Vide Tokenization

As we have seen earlier, the trust in the Blockchain platform which by itself is expected to act like a Disintermediating Trust machine must be above board with a due demonstration of credibility and high credentials. For this, the consortium leading the Blockchain platform's consensus mechanism, acting as the backbone for the entire solution should comprise of many leading stalwarts in the industry who could otherwise be competitors but are acting like collaborators in the Blockchain regime. The concept of Tokenization is a powerful concept that enables real-life assets and rights to be represented as digital equivalent value units on a Blockchain. Tokenization will enable members of different peer entities (that could be individuals or organizations) to represent their products and service in a common digital unit that could be traded. Ethereum pioneered the concept of decentralized applications that are powered by the respective Ethereum (native cryptocurrency of Ethereum Public Blockchain) compatible tokens which could be freely traded on the network. This will enable the launch of decentralized versions of all the real-life applications as we know like Amazon, Facebook powered by a Blockchain-based token.

LIBRA coin powered by a Blockchain platform spearheaded by Facebook is a precursor of things to come with respect to migration of today's centralized systems on to a Blockchain-based infrastructure. Blockchain platform-based applications like LIBRA (promoted by a group with Facebook as the leading member), Hash graph are combining ranks of leading players in the global business ecosystem and Fortune 100 members to form strong governing boards that could act like the 'Trusted Third Party' in a fool proof manner. LIBRA is a Permissioned Blockchain digital currency proposed by the American social media company Facebook, The project, currency and transactions are to be managed and cryptographically entrusted to the LIBRA Association, a membership of companies from payment, technology, telecommunication, online marketplace and venture capital, and nonprofits. (Wikipedia).

Comparison between Centralised & Decentralised Asset coins		
Type of value exchange unit	Centralised Banking & Marketplace systems	Blockchain based decentralised systems
National Currency	Indian Rupee	Central Bank Digital Currency over a DLT
Global Currency	US Dollar	Bitcoin
Marketplace Token	Amazon Paybalance	LIBRA coin
Closed loop Merchant token	Mall Food court cash card	DAPP Token

Table 1.1 Understanding LIBRA coin, JPM Coin and Bitcoin with analogies

Currently, the US Dollar is seen as the globally interoperable currency accepted by most nations. In the recent past, several countries are experimenting with the concept of leveraging the internet for speedy transfer of value considering the impending proliferation of IOT & Industrial IOT-led Home automation, Industrial automation, and Smart City projects across the world. There has been a strong need felt for a digital equivalent of the national currencies giving rise to the concept of Central bank digital currency (CBDC), also called digital fiat currency (a currency established as money by Government regulation or law). Central Bank Digital Currency is different from virtual currency and cryptocurrency, which are not issued by the state and lack the legal tender status declared by the Government. Various countries are already experimenting with the concept of CBDC and

it is considered a transitory step to the ultimate eventuality of a fully digitized currency with the added security measure offered by a Blockchain approach. According to the BIS, today some 70% of central banks are looking at CBDC, with most of them considering Blockchain as the underlying technology.

Future World of Blockchain Consortiums

Vision for a Blockchain powered future

Fig 1.17a Future of Cross Border-Cross Domain Consortiums with free-flowing Interoperability of identities & Assets across platforms

- The possibilities of Blockchain enabled future are exciting:

    1. Intra-enterprise networks for automating internal processes and collaboration across departments/and subsidiaries in case of conglomerated and Parent companies with subsidiaries.

    2. Each enterprise participating in different Blockchain consortiums depending on the domain of operation. Legacy applications and backend existing applications need to be integrated with Blockchain APIs of different networks leading to a large-scale Business Process Reengineering and elimination of intermediaries

    3. New types of cross-border Networks and Consortiums offering enhanced value to their customers and realising new economies of scale, simplified processes will come into place.

    4. A Paradigm shift will lead to a new world order of Consortium led approach & competitor collaboration to deliver products and services, Extensive usage of APIs, Smart contract led automation of business logic and speedy & trusted execution

    5. Huge demand for professionals who thrive in a multi-enterprise, multi-cultural environment with ability to think strategically, educate customers and promote complex solutions by focusing on the benefits and facilitate governance in an impartial & balanced manner, work alongside the Technology & Innovation teams to solve the problems of their clients and their customers. A reference architecture for integrating different systems incorporating IoT devices for connecting, communications and controlling the different facets of our life, leading to 'Internet of Everything' has been provided by Industrial Internet Consortium in their Whitepaper. The same is summarised in the following figure

Types of Integration Approaches of IoT & Blockchain for Smart city Applications
(Ref: Implementation aspect IIOT & Blockchain , An Industrial Internet Consortium White Paper)

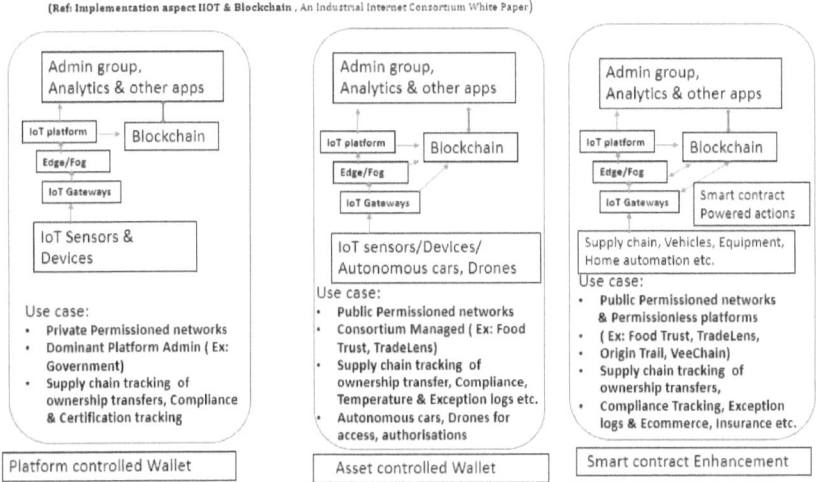

Fig 1.17b Blockchain Integration patterns for IoT based systems

Summary of 0–10 of Blockchain's Key Dimensions

The value proposition of Blockchain and the various key concepts can well be summarized in the following manner with a table of count from 0–10. These concepts summarize the many points we are covering across the book in appropriate contexts.

0. Single Points of Failure elimination to de-risk against cyber-attacks

1. Single Source of Truth for multiple transacting parties

2. Non-Trusting Parties can use Blockchain as a Trust Anchor instead of Humans—prevention of double spending

3. Triple Entry-Accounting with an indisputable and incorruptible record for tamper evidence

4. Key concepts for Business Blockchain: Shared ledger, Privacy, Smart Contracts, Consensus

5. Elements of Blockchain as per Gartner Framework: Encryption, Immutability, Tokenization, Decentralization, Distribution

6. Key application areas: Identity, Asset Provenance, Payments, Marketplace, Tokenization, Authenticity

7. DID benefits: Inclusion, Personalized engagement, Seamless Travel, Age and personal detail verification, Background checks, digital payments, One-click on-boarding

8. Consensus, Cooperation, Collaboration, Coordination, Cryptography (hashing, DSA, PKI, etc.), Confidentiality, Consortium, CBDC

9. Identity, Integration, Interoperability, Interchain, Intrachain, Interchange, initiate (Pay), Immutability, Intermediate removal 10. Step Methodology for Determining Blockchain Use case

10. Step methodology for determining applicability of Blockchain solutions as per WEF.

Blockchain Catalyzed Secured Automation with Trusted Transactions

Recently the world is witness to the uproar raised due to the failure on the part of Facebook to prevent the protection of identities of its members in the Cambridge Analytica case. In such a scenario, what then are the alternatives for global corporate citizens to express themselves freely and be connected across the world?

But then there is hope. The onset of the Artificial Intelligence empowered mammoth organizations with unlimited power, but with a single source of failure due to threats of cyber warfare, Ransomware and unscrupulous promoters have led to the people embracing Blockchain as a religion across the world. Blockchain has now given an unprecedented option for all peers across the world to be connected in a pseudonymous manner.

This is now leading to a movement that is going to take strength in the future as Blockchain paradigm cuts across the various sections of the world with a versatile, empowered, and ubiquitous ecosystem. The conventional social media will no more be the preferred choice of communication.

The implementation of Blockchain-enabled KYC and self-sovereign identity will empower the citizens across the world to own their own data and use the various platforms designed for specific purposes in a safe and secure way while monetizing their activities.

The data in the Blockchain-enabled future will put the power back into the hands of the individuals. The identities of the world's citizens will be protected by Blockchain as they operate in a pseudonymous manner with metadata visible to the advertisers and platforms.

They can give permission to the platforms and data consumers, while they get paid for the same as well as their social media contributions.

Several Blockchain-enabled platforms are already gaining traction and the same are listed in the book Blockchain the Untold Story, which examines the Blockchain technology in all its earnest for the benefit of a safe and secure future for global citizens.

Welcome to a world of 'Hope' and empowerment facilitated by this amazing paradigm of our era, Blockchain! The excitement to adopt this new paradigm can be visualized by the amount of activity that is taking place across the world from Companies, Consortia and Countries including the European Union and the United Nations.

"Trust, but verify" is a rhyming Russian proverb, used by Former USA President Ronald Reagan in the context of nuclear disarmament discussions with the Soviet Union. Blockchain with its ability to act as a Trust Anchor and provide an auditable track record closely symbolizes this mantra.

Blockchain Across Countries

Many Governments across the world are exploring/leveraging Blockchain to offer transparent, efficient, and cost-effective services to their citizens and industry. The following are some of the noted efforts by Governments across the world.

1. **Brazil** is using Blockchain-based Government e-Procurement to put a check on corruption in all Government purchases. Online Bid Solution, a Blockchain-based platform tracks the process of

public biddings for Government projects and purchase processes between various co-operative societies and industry bodies.

2. **China's** government has made blockchain a national priority in several directions, including the recently launched Blockchain Services Network #**BSN**), which is now being piloted in Chinese cities.

Between the CBDC and the BSN, blockchain technology has huge implications both for China's economy and perhaps international politics. China hopes to increase efficiency in its payments system and in how trade with its international partners is orchestrated. China is leveraging Blockchain to fight corruption and disintermediate tax collections. China's Government is launching its digital currency powered by Permissioned DLT and facilitate the Public Blockchain ecosystem in a big way. Leading companies like Alibaba, Tencent and thousands of other companies are working on implementing Blockchain projects across every possible use case, spurring innovation, and efficiency to a new plane.

China is implementing Blockchain in over 500 smart city projects and has encouraged over 1000 Start-ups to work on exploring the use cases in Blockchain technology through a government supported framework.

3. **Dubai** is patronizing Blockchain to eliminate all paper records across its governance, land records management, Police evidence tracking, Passport and VISA tracking, Cross/border remittances, Citizen Medical records tracking, etc., and save 5.5 billion dirhams annually in document processing alone equal to the one Burj Khalifa's worth of value every year.

Giving itself an ambitious target of becoming a paperless country, Dubai Government has been pioneering Blockchain and emerging technologies through a series of measures.

Dubai Land Department is employing the Blockchain in three initiatives (Ownership verification in DLD Mobile Application, Property sale by Developer and Smart Leasing Process) targeting the improvement of providing the services, improve the collaboration with other parties involved the real estate market and to create a secured digital assets A sample copy of a Land deed digitised on a Blockchain is given in the following figure.

شهادة ملكية عقار

Title Deed

Issue Date	11/12/2018	تاريخ الاصدار
Mortgage Status:	Not mortgaged غير مرشونة	حالة الرهن:
Property Type:	Villa فيلا	نوع العقار:
Community:	Al Yelayiss 2 2 البلايس	المنطقة:
Plot No:	178	رقم الأرض:
Building No:	TS HYT TH-V-451	رقم المبنى:
Area Sq Meter :	196.28	المساحة الكلية متر مربع :
Area Sq Feet :	2,112.74	المساحة الكلية بالقدم المربع :

Owners numbers and their shares:	Area (Sq Meter) \ المساحة بالمتر المربع	رقم و اسماء الملاك وحصصهم:
(5120709) RASHED AHMAD RASHED ALMULLA ALFALASI	196.28	(5120709) راشد احمد راشد الملا الفلاسي

Purchased from NSHAMA PROPERTIES OWNED BY NSHMI DEVELOPMENT ONE PERSON COMPANY L.L.C by the Land Registration No. : 225832/2015 Date 12/11/2018 for the amount 1226888 Dirham One Million and Two Hundred and Twenty Six Thousand and Eight Hundred and Eighty Eight Dirhams Only Dirhams

This property and its ownership is subject to the terms of the jointly owned property declaration of the above mentioned community and to the regulations issued in accordance with it as may be amended from time to time

Approved Signature توقيع معتمد

43266:2018

DUBAI LAND DEPARTMENT (565) دائرة الأراضي والأملاك

- Digital data of this certificate is securely stored on blockchain
- This certificate is electronically issued and no signature is required
- Any changes in the certificate make it void
- It is prohibited to hold this certificate by any other party

1 / 1

Fig 1.18 Blockchain record of a Land title deed in Dubai

Further, Dubai is one of the leading exponents of all emerging technologies.

Status of some of the Blockchain initiatives in Dubai are summarised in the following figure:

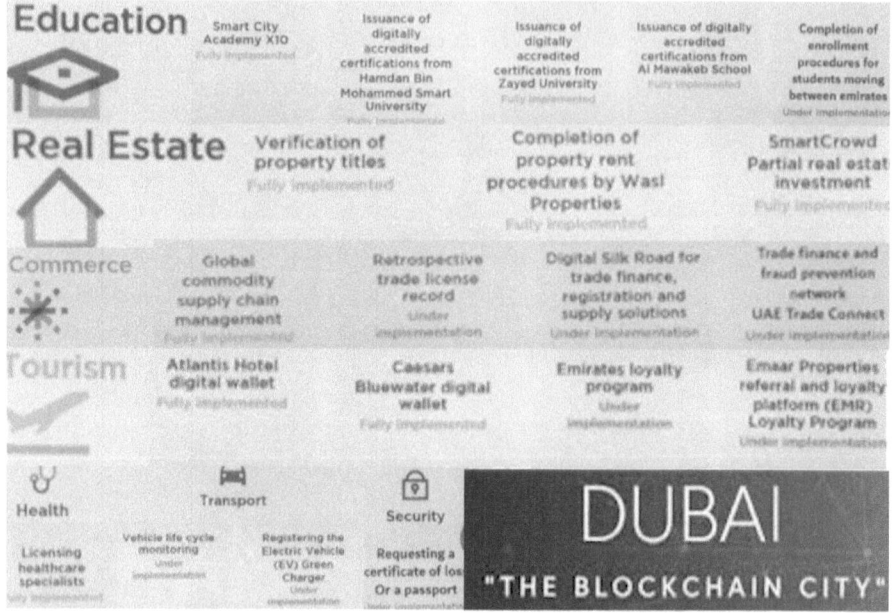

Fig 1.19 Blockchain projects being implemented with encouragement from Dubai Govt

4. **Estonia** a small country with 1.3 million population and an erstwhile part of Soviet Union, is extensively using Blockchain for the Integrity of data pertaining to all public and citizen records, Critical Infrastructure Protection and Secured access of all Government services to citizens through a Blockchain-enabled digital identity. Estonia secured all its citizens' medical records on a Blockchain.

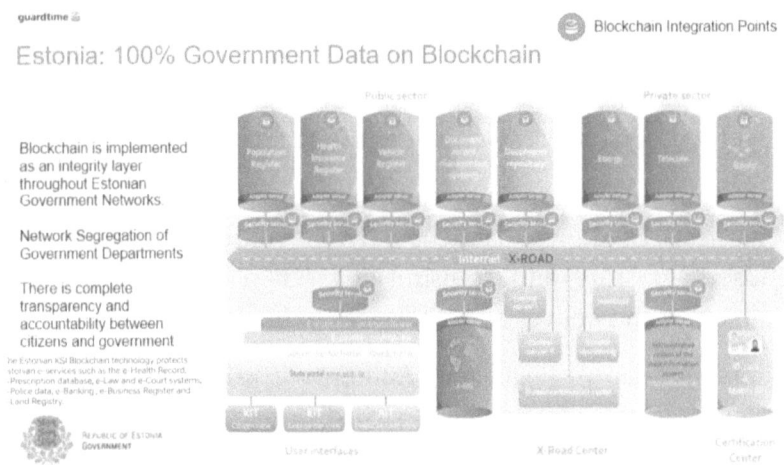

Fig 1.20 Estonia's integrated Blockchain platform to access citizen services

Source: Estonia Govt records in public domain

Estonia leverages KSI Blockchain technology to secure the Electronic Health Records (e-Health Record) of its 1.3 million citizens. The e-Health Record system offers an integrated view of the patient's records, test results including image files such as X-Rays issued by different hospitals and laboratories. The integrity of the records is ensured by the Blockchain while the doctors can get a complete view of the medical history of their patients via the e-Patient portal.

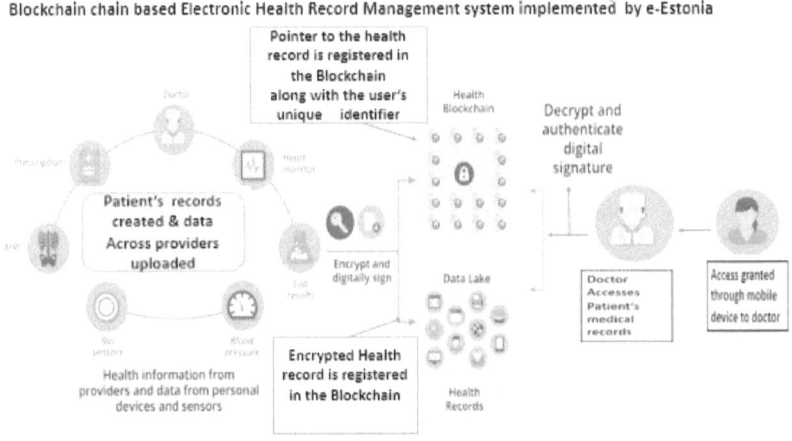

Fig 1.21 Integrated e-Health Record System of e-Estonia

The patients can have a complete record of their reports, transactions and doctor access and comments, maintain privacy and secrecy of their information while getting an integrated view of their personal records as per the access privileges built in.

5. **European Union:** European Blockchain services infrastructure project, EBSI, was launched by the European Union enables users to store and transmit data in a secure, decentralized manner and deliver better services to Europe's citizens.

6. **India's** Telecom Regulatory Authority is using Distributed Ledger Technology for tracking Unsolicited Commercial Communication. Several states, Ministries, Income Tax department, Customs department, Public Sector Undertakings, NPCI and Police departments are vigorously exploring Blockchain to improve transparency, efficiency and eliminate corruption and fake products, documents, identity, and certificates menace.

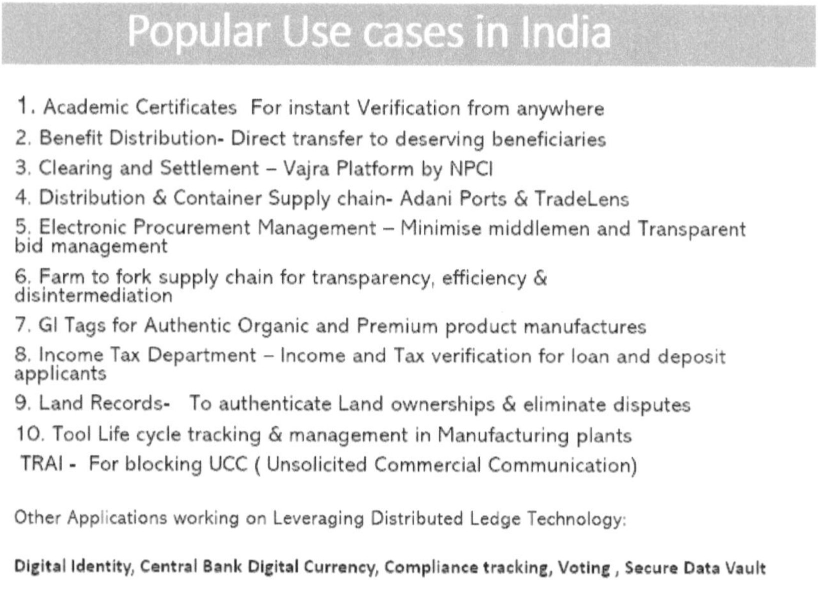

Popular Use cases in India

1. Academic Certificates For instant Verification from anywhere
2. Benefit Distribution- Direct transfer to deserving beneficiaries
3. Clearing and Settlement – Vajra Platform by NPCI
4. Distribution & Container Supply chain- Adani Ports & TradeLens
5. Electronic Procurement Management – Minimise middlemen and Transparent bid management
6. Farm to fork supply chain for transparency, efficiency & disintermediation
7. GI Tags for Authentic Organic and Premium product manufactures
8. Income Tax Department – Income and Tax verification for loan and deposit applicants
9. Land Records-   To authenticate Land ownerships & eliminate disputes
10. Tool Life cycle tracking & management in Manufacturing plants
TRAI - For blocking UCC (Unsolicited Commercial Communication)

Other Applications working on Leveraging Distributed Ledge Technology:

Digital Identity, Central Bank Digital Currency, Compliance tracking, Voting , Secure Data Vault

Fig 1.22 Popular Use cases in India

7. **Singapore** is working on a Blockchain-based payment system using digital Singapore Dollars, that can be used to execute inter-bank and cross currency remittances quickly and

affordably and with fewer intermediaries. Being a global hub for Finance and Supply Chain activities and organizations, Singapore Government enables a vibrant Blockchain ecosystem for enterprises to experiment and implement entire spectrum of Permissioned and Permissionless Blockchain applications across Finance, Supply Chain, trade finance, Crowdfunding, health insurance, Digital SGD, Academic certificates, etc.

8. **Uganda** is leveraging Blockchain in its Pharma Supply Chain to fight fake drug menace by eliminating them.

9. **UK** Government has been exploring Blockchain for several use cases like Central Bank Digital Currency for instant Inter-bank remittances, clearing and settlement, land records management, Government Data Provenance, Voting, Benefit and Charity distribution and Food safety in Supply Chains.

10. **USA:** US Government is working extensively on several Blockchain projects in Pharmaceuticals, Food, Cannabis, Defence Supply Chain provenance, health record tracking, Clinical records management, etc. Department of Homeland Security is researching Blockchain extensively for Critical Infrastructure protection using Blockchain enabled identification systems.

11. **Thailand:** Thailand Government is extensively experimenting with Blockchain for a variety of applications involving Digital identity, Supply chain and Central Bank Digital currencies.

The growth of Consortia in the recent past in furthering the implementation and adoption of the Blockchain paradigm cannot be overstated.

Blockchain Consortiums

1. **Enterprise Ethereum Alliance** (https://entethalliance.org/) The EEA is a member-led no-profit industry consortium consisting of over 150 organizations to work around the adoption of Ethereum Blockchain technology as an open-standard to

empower enterprises working for a decentralized world across various use cases for Government and enterprises. It comprises global leading organizations including ConsenSys, CME Group, Cornell University's research group, Toyota Research Institute, Samsung SDS, Microsoft, Intel, J. P. Morgan, Cooley LLP, Merck KGaA, DTCC, Deloitte, Accenture, Banco Santander, BNY Mellon, ING, National Bank of Canada MasterCard, Cisco Systems, Sberbank and Scotiabank.

2. **Hyperledger** **(https://www.hyperledger.org/)** Hyperledger (or the Hyperledger project) is a cross sectoral umbrella project of open-source Blockchains and related tools, started in December 2015 by the Linux Foundation,[1] and has received contributions from IBM, Intel and SAP Ariba, to support the collaborative development of Blockchain-based distributed ledgers. Global leaders across various domains like Technology platforms, Blockchain platforms, System integrators, Management consultancies, Financial services, Manufacturing and Supply Chain organizations collaborate as part of the Hyperledger foundation on various solutions for different applications.

3. **R3** (https://www.r3.com/) led consortium R3 (R3 LLC) is a New York-based Enterprise Blockchain technology company leading an ecosystem of more than 300 firms working together to build distributed applications on top of Corda (known as CorDapps) for usage across industries such as financial services, insurance, healthcare, trade finance, and digital assets. Corda, an open-sourced Distributed Ledger Platform records, manages and synchronizes financial agreements and standardizes data and business processes.

4. **TradeLens** https://www.tradelens.com TradeLens, jointly developed by IBM and Maersk GTD is an open and neutral Blockchain platform serving the ecosystem of the supply chain in information sharing and collaboration across the value chain, thereby increasing industry innovation, reducing trade friction and ultimately promoting more global trade.

5. **IBM Food Trust:** https://www.ibm.com/in-en/Blockchain/ solutions/food-trust IBM Food Trust is a consortium of producers, suppliers, manufacturers, retailers and any other business serving the global food supply chain with a focus to create a trusted, transparent, safer and efficient food system across the world. Walmart is one of the key members of the consortium along with IBM.

6. **MediLedger:** https://www.mediledger.com The MediLedger project is a collaboration between Chronicled and The LinkLab, connecting leading pharmaceutical manufacturers and distributors to explore Blockchain technologies by bringing together expertise in both pharmaceutical supply chain and Blockchain technologies. The project is building an industry owned Permissioned Blockchain network for the pharmaceutical sector based on open standards and specifications. Network nodes are set to be distributed among and operated by industry participants and technology providers who serve the industry with an aim to meet compliance and regulatory requirements of the industry, provide Track-and-Trace for the products and eliminate the role of fake drugs and also to enable the parties for easy mutual reconciliation of accounts.

7. **Trusted IOT Alliance:** https://www.iiconsortium.org The Trusted IOT Alliance is an open-source software foundation with a mission—to leverage the power of Blockchain and Distributed Ledger Technologies. It consists of leading IOT companies in the world who are together working to unravel the disruptive power of all emerging technologies. It is now consolidated with The Industrial Internet Consortium, a global not-for-profit partnership of Industry, Government and Academia.

8. **MOBI:** https://dlt.mobi The Mobility Open Blockchain Initiative is a consortium formed by global automobile sector majors across the world to leverage Blockchain technologies for safer and greener automation-powered transportation technologies. Almost all the Fortune 500 members from the automobile

technology, vehicle and ancillary manufacturing, consultancy and services space are members of the consortium.

9. **B3I:** https://b3i.tech/home.html B3I is an insurance industry consortium and comprises of over 20 insurance majors and 20 customers and service providers. The consortium is focusing on leveraging Blockchain to improve coordination among the players and eliminate fraud. It is working on developing standards, protocols and network infrastructure to remove friction in risk transfer and give end consumers of insurance better and faster access to insurance.

10. **Synaptic Health Alliance** (https://www. synaptichealthalliance. com/) Synaptic Health Alliance is a consortium of health care majors like Aetna, Humana, MultiPlan, Optum, Quest Diagnostics, and United Healthcare. It is working to leverage Blockchain to tackle the problems associated with data duplication across multiple silos that involve feeding activities such as claims processing, payment integrity processes, provider and member attribution, provider directories, and more.

11. **Global Shipping Business Network:** https://www.cargosmart. ai/en GSBN is a global consortium of leading global shipping organizations and port operators formed to leverage the power of emerging technologies, especially Blockchain for smooth cargo transportation through seamless documentation, consignment traceability and financial transactions including support to Trade Finance activities. The members include industry-leading ocean carriers and terminal operators, namely CMA CGM, COSCO SHIPPING LINES, COSCO SHIPPING Ports, Hapag-Lloyd, Hutchison Ports, OOCL, Port of Qingdao, PSA International, and Shanghai International Port Group (SIPG), There are many such leading consortia as mentioned in the table providing the list of Blockchain consortia and this is a growing trend that promises to catalyse the migration of all enterprises in every sector to leverage Blockchain for collaboration for seamless service to customers and increased productivity. The Governance of the Consortia needs to be fair,

transparent and democratic. In future, the consortia will be leading providers of jobs across the emerging technology space unlike today's centralized organizations

Summary

The Chapter throws light on several dimensions of Blockchain Technology like:

- Concepts behind Blockchain Technology and its utility to enterprises and Government.

- Workings of different Blockchain platforms like Bitcoin, Ethereum, Hyperledger and R3 Corda

- Different Use cases of Blockchain across Government and Enterprises.

- Different Use cases being explored in India

- How Blockchain is being adopted in different countries across the World.

- Concept of Blockchain Consortiums and have read about different types of Consortiums operating across different domains, globally.

CHAPTER 2

Internet of Things – A key Building Block of Smart Cities

Introduction

The Internet of things is a system of interrelated computing devices, mechanical and digital machines provided with unique identifiers and the ability to transfer data over a network without requiring human-to-human or human-to-computer interaction.

With the help of different types of Sensors connected to the internet and transmitting various types of information, Internet has revolutionised the way we interact with our surroundings, environment and our devices. With 5G about to be launched across the world, there is going to be a big explosion of data and information traffic across all connected beings and things, leading to several new paradigms affecting our life, mostly for the better and partly for the worse as far as cyber security is concerned, In this chapter, we examines the various elements of IoT ecosystem and understand the various applications made possible with the interconnections made possible by the Internet of Everything.

Machine generated data derived from the phenomenal growth in the number of sensors and machines used to measure and record the events and situations in the physical world. The output of these sensors is machine-generated data, and from simple sensor records to complex computer logs, it is well structured. As sensors proliferate and data volumes grow, it is becoming an increasingly important component of the information stored and processed by many businesses. Its well-structured nature is suitable for computer processing, but its size and speed are beyond traditional approaches. The various sources of IoT data are depicted in the following section.

Data from Sensors

a. **Fixed sensors**

- Home automation

- Weather/pollution sensors

- Traffic sensors/webcam

- Scientific sensors

- Security/surveillance videos/images B. Mobile sensors (tracking)

- Mobile phone location

- Cars

- Satellite images

b. **Data from computer systems**

- Logs

- Web logs

The amount of data transacted through the internet grows manifold as the communication between the devices and their principals grows.

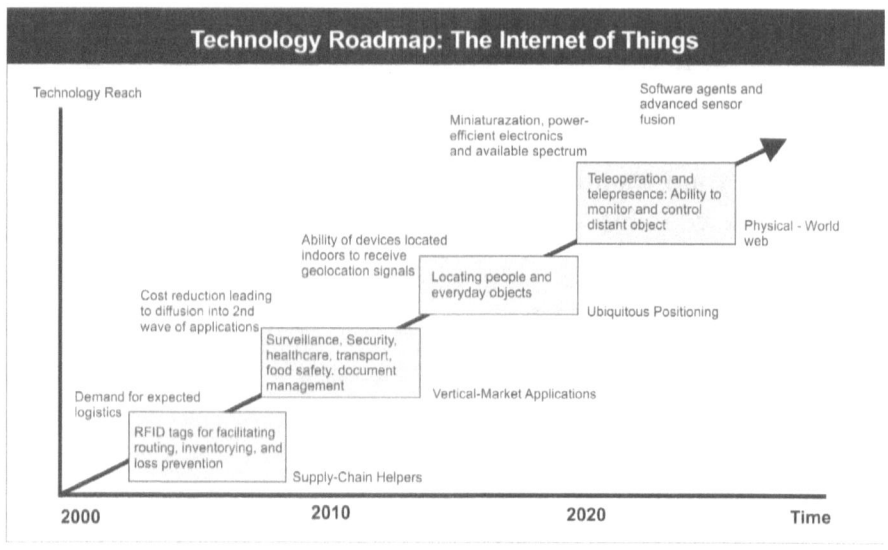

Fig 2.1 IoT Features and Benefits across ages

Starting with RFID (Radio frequency identification) chips in the Year 2000, the advances in the field of Internet of Things are expected to permeate to the entire world connecting humongous number of devices with sensor that are then hooked up to the internet through advanced communication platforms and technologies. By 2020, the number of interconnected devices and the associated business opportunity is expected to grow manifold to over 30 Billion connected objects and 50 Trillion US Dollars in revenue opportunity.

IoT will impact all areas cutting across consumer and industrial usage. While smart home, wearables and personal trackers are good use cases for Consumers, IoT can be leveraged across all areas of design, manufacturing, storage, transportation, maintenance and live performance tracking through the value addition cycle of products and services to dramatically enhance the productivity and profitability of organizations.

Components of IoT System

Let us examine the major components of IoT:

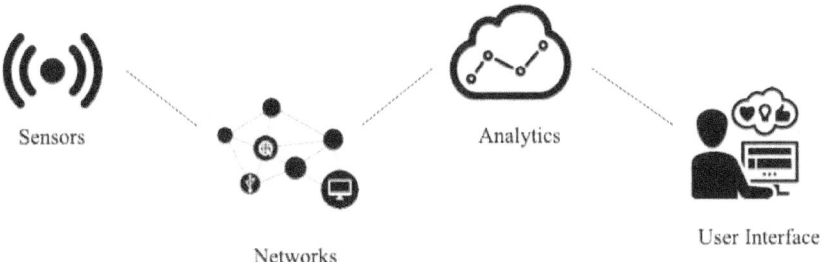

Fig 2.2 Major Components of IoT

1. SENSORS AND SMART DEVICES

According to IEEE, sensor is an electronic device that produces electrical, optical, or digital data derived from a physical condition or event. Smart devices are devices with one or more sensors that can bundle together to do more than just sense things. E.g. Smart phones have multiple sensors such as GPS, accelerometer, camera.

These sensors or smart devices are used to pick and collect data from the environment which is the most rudimentary step in IoT implementation.

2. NETWORKS

Networks are the components which enable connectivity in the IoT system. The second step of the implementation is to transmit the signals collected by the sensors through one of the various mediums of communications such as cellular networks, Wi-Fi, Bluetooth, Low Power Wi-Fi, Wi-Max, regular Ethernet, etc. to all the components of the network. Low power, low cost wireless transmitting devices are preferred due to their long battery life and efficiency.

Since Internet of things creates huge amounts of data from devices, cloud infrastructure is leveraged upon to process, manage and store data in real time. Cloud system integrates billions of devices, sensors, gateways, protocols, data storage and provides predictive analytics, and that brings us to the next component – Data processing and Analytics.

3. ANALYTICS

The third step in IoT implementation is extracting insights from data for Analysis. The analysis can range from something very simple, such as checking if the temperature reading on devices such as AC or heaters is within an acceptable range to something complex, such as identifying objects using computer vision on video. Information is very significant in any business model and predictive analysis ensures success in concerned area of business line.

4. APPLICATIONS AND USER INTERFACE

Applications refer to the various activities performed using the IoT networks and are activated through a User interface. The application layer is a user-centric layer which executes various tasks for the users. There exist diverse IoT applications, which include smart transportation, smart home, personal care, healthcare, etc.

Consider a scenario that the analysis from the previous step has shown that there is an intruder in the house. So, what next? This information must be made available to the user in some way by triggering alarms or through notifications via phone, text or email.

Also, an interface may be required to enable users to actively monitor the status of their IoT system. Ideally, the user interface is well designed to support minimum effort for users and encourage more interactions.

IoT Layers

Cisco, the industry leader in IoT platforms, offers an integrated solution for the management of IoT devices. The Cisco IoT/M2M architecture is composed of four layers, some are similar to those described in conventional Cisco network architectures.

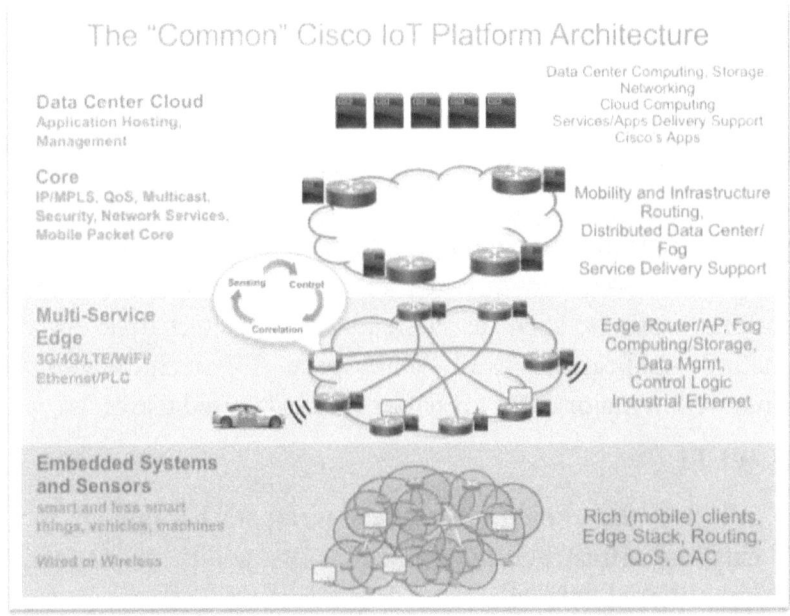

Fig 2.3 IoT/M2M Network Architecture Layers

Source: https://www.cisco.com/c/en/us/about/security-center/ secure-iot-proposed-framework.html

Internet of Things covers a huge range of industries and use cases across cross-platform deployments and cloud systems in real-time. Hence, there exist certain standards and protocols to allow the communications between the devices and servers in more interconnected ways. The major IoT protocols at various layers in the IoT Architecture stack are depicted in the following graphic (Source: Simon Ford – Director of IoT Platforms ARM).

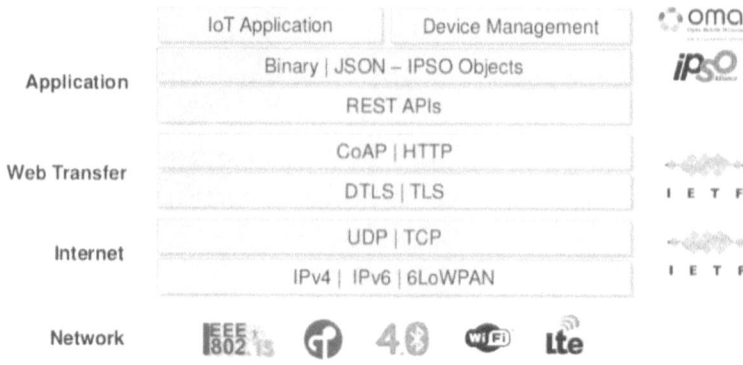

Fig 2.4 Major IoT Protocols at Various layers (Simon Ford – Director of IoT Platforms ARM)

Many communication technologies such as Wi-Fi, Bluetooth, ZigBee and cellular are well known, but there are also several new emerging networking options such as Thread as an alternative for home automation applications, and Whitespace TV technologies being implemented in major cities for wider area IoT-based use cases.

1. **WI-FI**

   Wi-Fi is the choice for many developers because of its wide existing infrastructure as well as its ability to handle high quantities of data with rapid data transfer.

2. **BLUETOOTH**

   Bluetooth, a short-range communications technology is expected to be key for wearable products in particular, again connecting to the IoT via a smartphone in most of the cases. The new Bluetooth Low-Energy (BLE) – or Bluetooth Smart is a significant protocol for IoT applications as it offers significantly reduced power consumption.

3. **ZIGBEE**

   ZigBee, like Bluetooth, has a large installed base of operation, more in industrial settings. ZigBee/RF4CE offers low-power operation, high security, robustness and high scalability with high node counts and is well positioned to take advantage of wireless control and sensor networks in M2M and IoT applications.

4. CELLULAR

IoT applications requiring operations over longer distances can leverage GSM/3G/4G cellular communication capabilities. Cellular is capable of sending high quantities of data, especially for 4G, however the expense and power consumption is deemed to be high for many applications. Cellular can be ideal for sensor-based low-bandwidth-data projects that will send very low amounts of data over the Internet.

5. 6LOWPAN

6LowPAN (IPv6 Low-power wireless Personal Area Network) is a network protocol that defines encapsulation and header compression mechanisms.

6. Thread

Thread is a relatively new IP-based IPv6 networking protocol aimed at the home automation environment. It is primarily designed as a complement to Wi-Fi as it recognises that while Wi-Fi is good for many consumer devices that it has limitations for use in a home automation setup.

7. LoRaWAN

LoRaWAN targets wide-area network (WAN) applications and is designed to provide low-power WANs with features specifically needed to support low-cost mobile secure bi-directional communication in IoT, M2M and smart city and industrial applications. LoRaWAN is optimized for low-power consumption and can support large networks with millions and millions of devices.

Protocol	Standard	Frequency	Range	Data Rates
Wi-Fi	802.11n	2.4GHz and 5GHz	Approximately 50m	600 Mbps maximum, but 150-200Mbps is more typical.

Protocol	Standard	Frequency	Range	Data Rates
Bluetooth	Bluetooth 4.2 core specification	2.4GHz (ISM)	50-150m (Smart/BLE)	1Mbps (Smart/BLE)
ZigBee	ZigBee 3.0 based on IEEE802.15.4	2.4GHz	10-100m	250kbps
Cellular	GSM/GPRS/ EDGE (2G), UMTS/HSPA (3G), LTE (4G)	900/1800/ 1900/2100MHz	35km max for GSM; 200km max for HSPA	35-170kps (GPRS), 120-384kbps (EDGE), 384Kbps-2Mbps (UMTS), 600kbps-10Mbps (HSPA), 3-10Mbps (LTE)
6LowPAN	RFC6282	Bluetooth Smart (2.4GHz) or ZigBee or low-power RF (sub-1GHz)	N/A	N/A
Thread	Thread, based on IEEE802.15.4 and 6LowPAN	2.4GHz (ISM)	N/A	N/A
LoRaWAN	LoRaWAN	Various	2-5km (urban environment), 15km (suburban environment)	0.3-50 kbps

Table 2.1 Different IOT Protocols and their parameters

IoT will impact all areas cutting across consumer and industrial usage. While smart home, wearables and personal trackers are good use cases for Consumers, IoT can be leveraged across all areas of design, manufacturing, storage, transportation, maintenance and live performance tracking through the value addition cycle of products and services to dramatically enhance the productivity and profitability of organizations.

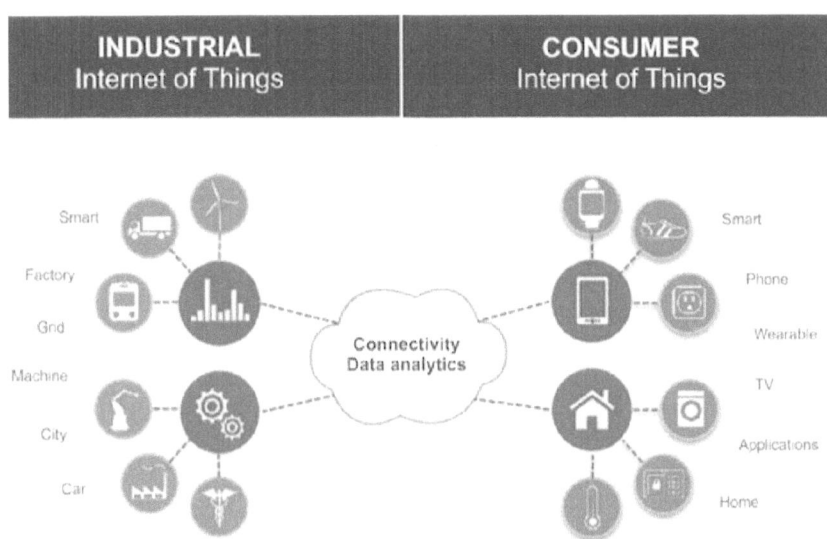

Fig 2.5 IoT for individual and Industrial use cases

As of now, IoT is extensively used in the following areas:

- Tagging and monitoring the movements and performance parameters of automobiles, animals, accessories and dependents.

- Information collection and dissemination for optimum agricultural output.

- Automatic meter reading and energy utilization management for all electrical devices.

- Security and Surveillance systems for a wide variety of establishments and connected homes

- Building automation and maintenance services.

- Machine to Machine connectivity offering communication, ecommerce and collaboration opportunity through high levels of Automation.

- Smart cities to effectively manage delivery of efficient products and service across every area that touches a citizen's life.

- Telemedicine and healthcare for remotely measuring and tracking the health parameters to offer quick resolution to patients in remote areas.

IoT – A Key Building Block of Smart Cities

Centralized platforms can offer innumerable services through cloud-based platforms by converting every dumb object into intelligent and communicable systems. Governments across the world have realized the potential of IoT and are implementing various projects under the category of 'Smart Cities,' while service, technology and infrastructure providers are leveraging advanced technologies and innovation to dramatically optimize and enhance the capability and productivity of systems across every field we can think of.

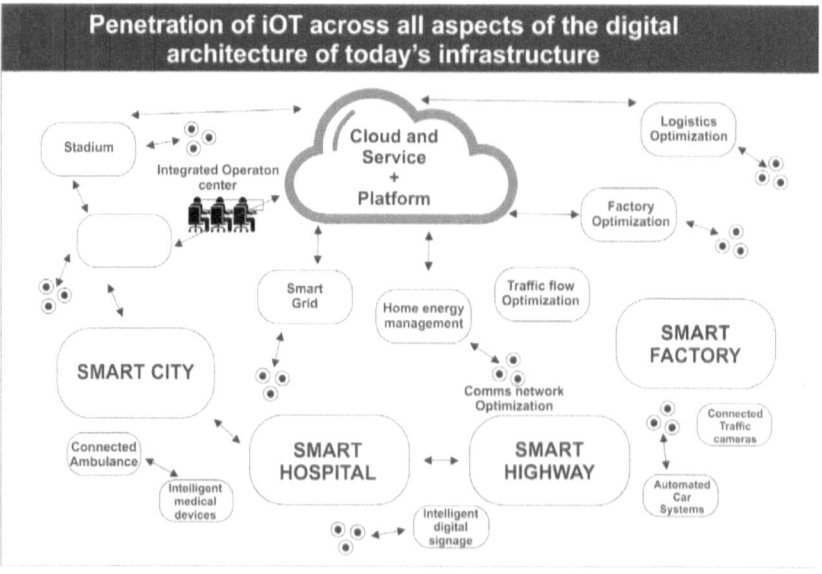

Fig 2.6 IoT powering Smart City infrastructure and Applications

The Internet of Things is a new paradigm that is expected to make life much more convenient for citizens across the world. With things connected to the internet, their users can access and control it from anywhere. This dramatically increases the productivity of human beings while creating huge opportunity for Data collection, analysis, monitoring, track ability, market opportunity.

Thereby IoT can create value for proactive organizations, by giving them a huge competitive advantage. This could also be a source of risk due to the vulnerability of these devices to malicious attacks by cyber terrorists.

Big Data technologies could play a crucial role in managing the humongous amount of information generated by the sensors deployed on all things that need to be monitored and managed.

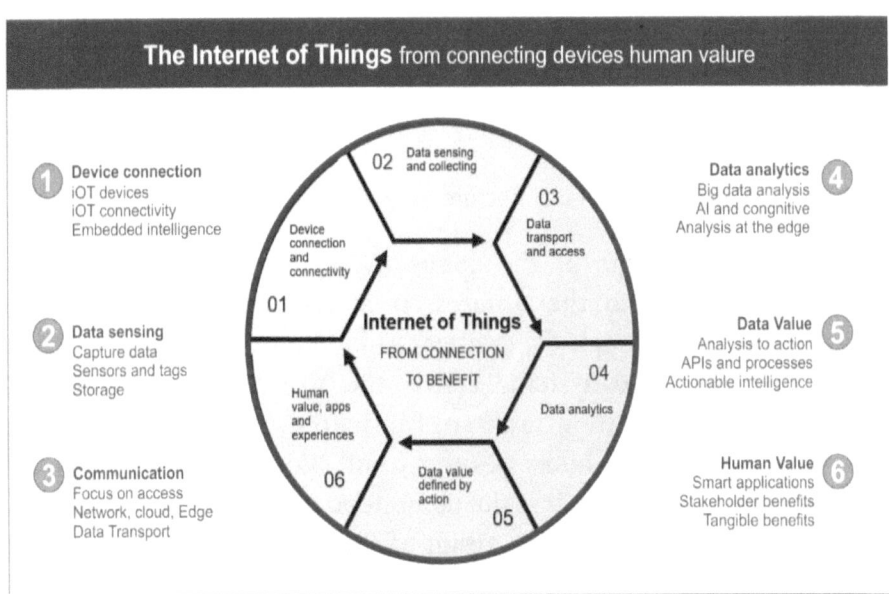

Fig 2.7 Data connectivity, Collection, Collation, Processing and Value Unlocking by leveraging IoT

Companies like General Electric, Rolls Royce, Shell and many more are proactively using the IoT technology to dramatically improve their efficiency as well as reduce the chance of failures and defects by combining the technology with cutting edge tools facilitated by

Analytics, Artificial Intelligence and Machine Learning, as depicted in the table enumerating the case studies on Big Data Analytics.

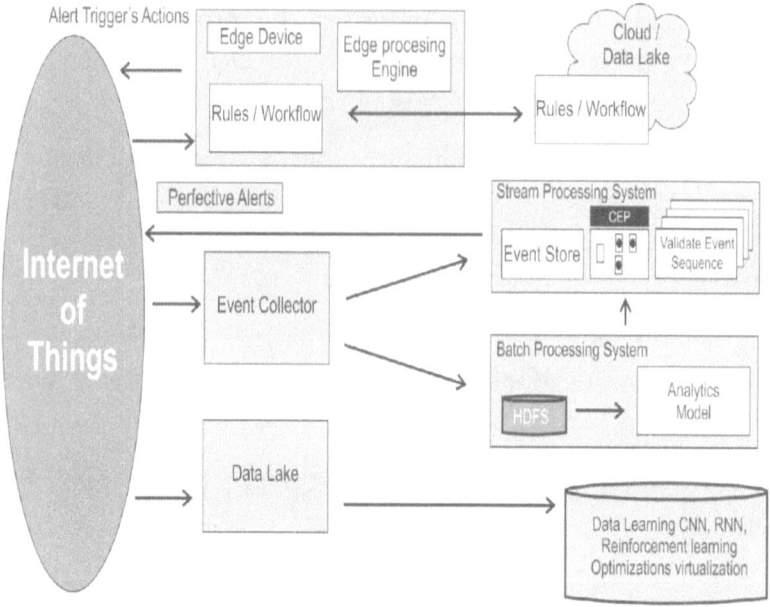

Fig 2.8 Typical architecture of IoT and its integrations

As we take advantage of the connectivity revolution, we are also exposing ourselves to the unforeseen attacks on the IoT systems across the world through cyber-attacks like the October 2016, DDOS attack that paralyzed systems across the world. https://www.cnbc.com/2016/10/22/ddos-attack-sophisticated-highly-distributed-involved-millions-of-ip-addresses-dyn.html. Blockchain offers a great de-risking mechanism against the ubiquitous Internet connectivity and infinite attack surface that is a result of the IOT penetration & Smart City explosion.

Summary

This chapter throws light on various aspects of IoT like

1. Definition of IoT and the possibilities it brings forth

2. Benefits of IoT

3. Areas & domains impacted by IoT along with applications

4. Components of IoT system

5. Layers of IoT system

6. Integration of IoT with other data collections, analysis and presentation layers

7. IoT and the Smart city connection

CHAPTER 3

Smart Cities – The Future of Urbanisation

Introduction

"The rudiments of what constitutes a Smart Sustainable City which we define as a city in which ICT is merged with traditional infrastructures, coordinated and integrated using new digital technologies." (Batty, et.al, 2012) A Smarter City uses technology to transform its core systems and optimize finite resources. At the highest levels of maturity, a Smarter City is a knowledge-based system that provides real-time insights to stakeholders, as well as enabling decision-makers to proactively manage the city's subsystems. Effective information management is at the heart of this capability, and integration and analytics are the key enablers. (IBM, 2013)

The concept of Smart city as a symbol of Sustainable Urbanization has been gaining traction across the world. Through Retrofitting of existing areas, Redevelopment of dilapidated areas or through completely green field projects, the countries are seeking to develop new infrastructure with the aid of advanced technologies and modern design methodologies for a sustainable future. In this chapter, we take a look at the concept of Smart cities and the various facets of the Smart cities. This concept holds the beacon light for a future of hope, sustainability and balanced & enjoyable lifestyle for the global citizens.

Importance of Sustainable Urbanization

The population of the world has grown from under 1.5 billion to over 7 billion over the past 100 years. This is putting enormous pressure on the resources like water, land, food, breathing air, energy resources and opportunities for people with increased competition while, automation is leading to reduction in manpower requirement to perform routine jobs.

There has been an accelerating migration to the urban areas from rural areas, leading to increased population densities of unlimited levels while, there is lowered availability of manpower to undertake rural activities like farming and animal husbandry that feed the growing population. There is an urgent need to leverage technology to manage this situation, work on the rural-urban imbalances and work on a sustainable environment.

This led to the concept of Digital Nations and Smart Cities. While Digital Nations work on leveraging the technological developments to enhance the all-round wellbeing of their citizens with improved quality of life, Smart cities are developed and promoted across the geography of nations to reduce the stress on the urban settlements.

China is developing over 500 smart cities while India is developing over 150 smart cities that are aimed at reducing the load on the top 10 metros in their respective countries. The smart cities must be developed with a balance across multiple dimensions like governance, healthcare, recreation, education, social & cultural, employment opportunities, citizen security while above all, maintaining the environmental sustainability. Technology & connected things, systems and people form an important component of the Smart City concept.

Defining a Smart City

Major technological, economic and environmental changes have generated increased interest in smart cities, including climate change, economic restructuring, coronavirus, the move to online retail and entertainment, ageing populations, urban population growth and pressures on public finances

A smart city is an urban area that uses different types of electronic Internet of things (IoT) sensors to collect data and then uses insights gained from that data to manage assets, resources and services efficiently, in return using that data to better improve the operations across the city. (Wikipedia).

This includes data collected from citizens, devices, buildings and assets that is processed and analysed to monitor and manage traffic and transportation systems, power plants, utilities, water supply networks, waste management, crime detection, information systems, schools, libraries, hospitals, and other community services.

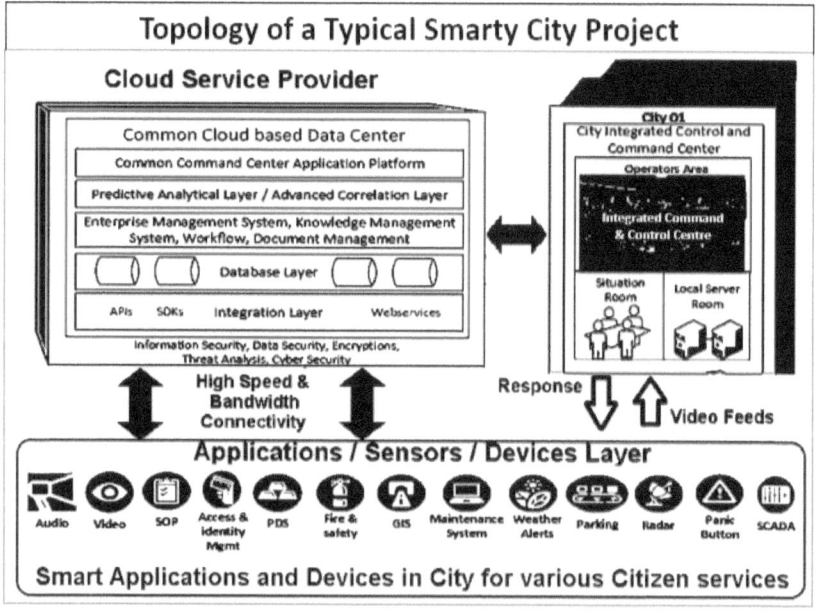

Fig 3.1a 3 Major components of Smart city It infrastructure (Source: Indian Govt records)

The smart city concept integrates information and communication technology (ICT), and various physical devices connected to the IoT network to optimize the efficiency of city operations and services and connect to citizens.

One of the key components of the Smart city concept is ICCC (or Integrated Command and Control Centre) that monitors the information collected and collated from various sensors and communication devices deployed for a variety of applications.

The various surveillance benefits of ICCC are given in the following table.

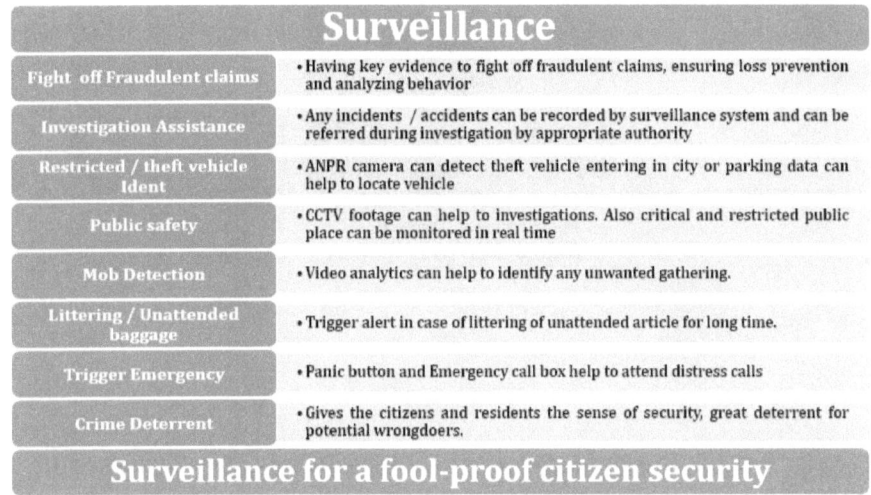

Surveillance	
Fight off Fraudulent claims	• Having key evidence to fight off fraudulent claims, ensuring loss prevention and analyzing behavior
Investigation Assistance	• Any incidents / accidents can be recorded by surveillance system and can be referred during investigation by appropriate authority
Restricted / theft vehicle Ident	• ANPR camera can detect theft vehicle entering in city or parking data can help to locate vehicle
Public safety	• CCTV footage can help to investigations. Also critical and restricted public place can be monitored in real time
Mob Detection	• Video analytics can help to identify any unwanted gathering.
Littering / Unattended baggage	• Trigger alert in case of littering of unattended article for long time.
Trigger Emergency	• Panic button and Emergency call box help to attend distress calls
Crime Deterrent	• Gives the citizens and residents the sense of security, great deterrent for potential wrongdoers.
Surveillance for a fool-proof citizen security	

Fig 3.1b Surveillance benefits of Integrated Command and Control Centre

The ICCC will be used to:

- Connect, collect data, and monitor sensors and cameras across the city to have a continuous ground level activity status related information

- Keep in continuous touch with the officials on ground to solve customer issues

- Create, Implement & Update fine-grained event – response protocols across scenarios

- Increase collaboration across various departments inside and outside government.

- Facilitate data driven decision making always and at all levels of Governance.

Fig 3.2 The Integration of all ICT infrastructure to a Central monitoring system (ICCC)

- Smart city technology allows city officials to interact directly with both community and city infrastructure and to monitor what is happening in the city and how the city is evolving.

- ICT is used to enhance quality, performance and interactivity of urban services, to reduce costs and resource consumption and to increase contact between citizens and government.

ICCC Architecture

A holistic ICCC platform will basically include four layers in the system, as mentioned below:

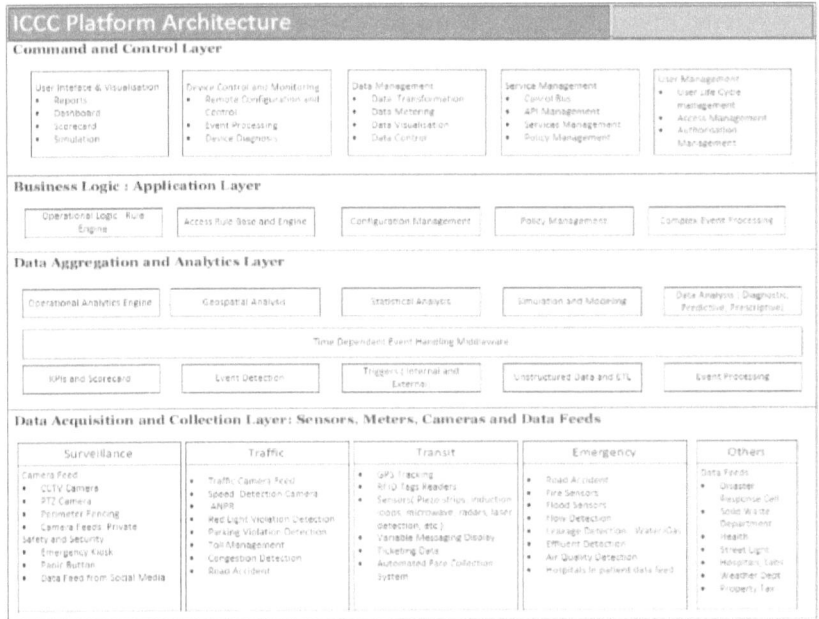

Fig 3.3 Integrated Command and Control Centre – Platform Architecture

(Source: Ministry of Housing & Urban Affairs (MoHUA), Govt of India Records in public domain)

Data Acquisition Layer: This layer collects real time data from sensors devices, data sources, static and real time data feeds from different applications, systems and databases etc. for air and water quality monitoring, ambient light sensors for street light management, metering devices, telematics and location based devices, proximity sensors, surveillance and safety cameras, sensors for disaster detection, level sensors for solid waste management etc.

An ICCC can either collect data from sensors or process the same to generate information from the data collected and aggregated through its various components to generate alerts, or it can connect to COTS (Commercial off-the-shelf) and bespoke applications so that alerts are generated by the integrated COTS/bespoke application/systems.

This layer enables other components of ICCC to aggregate, consume and process the data for deriving information.

Data Aggregation and Analytics Layer: This is responsible for deriving information and intelligence from data captured from various data sources through data acquisition layer. Data Aggregation and Analysis comprises of components for extraction and transformation of data from different systems, data sources and different data formats. For ex: Health records are captured from Integrated Hospital Management System, traffic information is captured from Adaptive Traffic Management System and Ambulance could be tracked using Vehicle Tracking system in different formats. ICCC Data aggregation and Analysis Layer can process the information and allows users to use information from different systems as per requirements.

Data Analytics components are used to perform data churning to derive intelligence from different data sets across the domain. This intelligence can then be used for exception handling and visualization in different scenarios through various analysis using ICCC components or third-party tools/applications:

a) Predictive Analytics b) Diagnostic Analytics c) Prescriptive Analytics d) Sentiment Analytics e) Video Analytics

This layer enables ICCC to derive intelligence from the information collected from Data Acquisition and Collection Layer.

Business Logic Application Layer: This is the core engine of the ICCC platform which help end user to design and configure standard operating procedure, manage external and internal trigger, policy implementation, and handling complex events. Application layer also helps ICCC to handle events in real time complimenting it with intelligence and information from various systems. Application layer also manages the response in different situations as per configured business logic.

Smart city applications are developed to manage urban flows and allow for real-time responses thus differentiating from those with a simple "transactional" relationship with its citizens (Source: Wikipedia)

The various services offered by Smart City Governments across domains to their subjects like Citizens, Businesses, Start-Ups are summarised in the diagram that follows.

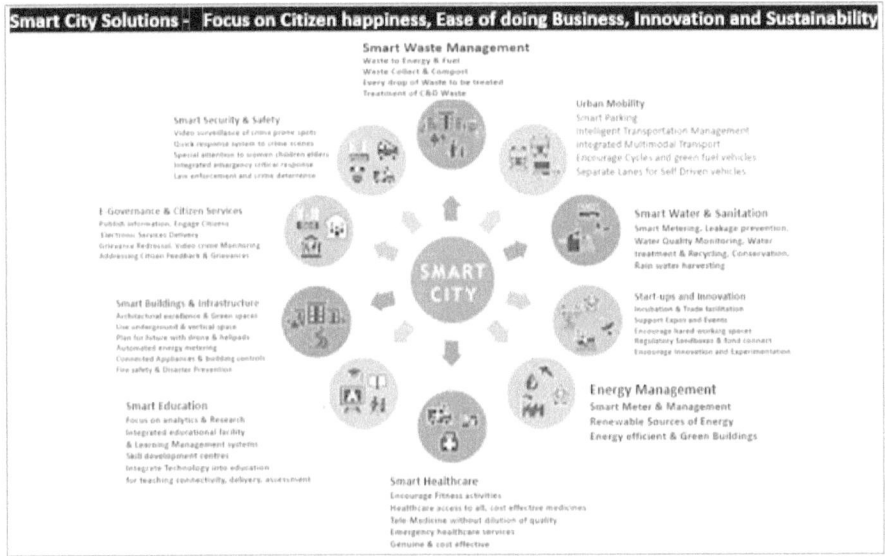

Fig 3.4 SMART City Solutions –Comprehensive portfolio of services to stakeholders

The various Smart services offered by the Smart city Governments by leveraging Smart Infrastructure are described in the following.

Smart Infrastructure

Smart City infrastructure consolidates the various themes of the Smart city for an all-round balanced development. This includes smart people, smart mobility, smart economy, smart living, smart governance, and smart environment. All these are interrelated and managed holistically through an integrated technology platform that collects data from various sensors employed in the respective domains, collect and transmit data for analysis and better decision making to improve citizens' lives. These different pillars of the Smart infrastructure and examined in this section.

Smart buildings: A smart building offers optimized and efficient management of all its physical systems as follows (as per UNCTAD):

a. Improve building energy efficiency (can improve by over 40%),

b. Reduce hazardous waste & increase easy waste recycling,

c. Ensure an optimum usage of water (by 30%), with operational effectiveness and

d. Reduce maintenance costs up to 30% & increase occupant satisfaction.

It is estimated that implementing smart building solutions could save as much as 30 per cent of water usage and 40 per cent of energy usage and reduce overall building maintenance costs by 10 to 30 per cent.

Smart mobility: Smart mobility facilitates quality transportation that is free of congestion by offering faster, greener, safer, convenient, accessible, predictable, and cheaper transportation options by leveraging data collected from a variety of sources about mobility patterns. A well interconnected multimodal Integrated Mobility System (IMS) built as a layer over network of feeders and integrated transportation fleet system to empower the citizen with real time information and increase the convenience of public transport, enables a substantial reduction of personalised private transport in favour of shared public transport reducing congestion & pollution.

This includes a variety of complimentary activities like:

a. Multimodal transport by combining mass transit systems effectively,

b. Energy efficient individual mobility systems that feature bicycles and electric vehicle,

c. Encourage sharing-ride sharing (or carpooling), vehicle sharing and, more recently, on-demand transportation.

d. Leverage various technology components including network of sensors, global positioning system-tracked public transportation, dynamic traffic lights, passenger information panels, automatic vehicle registration plate readers, closed-circuit television systems, navigation facilities, signalling systems and, most importantly, the capability of integrating live data from most of these sources.

Let us see how one of the leading global Smart cities implemented Smart mobility, solutions: A global top 10 ranked smart cities implemented the following activities:

i. Promoted car sharing and public transport: Redesigned public transport routes, express bus lines.

ii. Traffic monitoring and control system: All traffic lights across the city were combined into one centrally managed network.

iii. Two mobile applications: Developed mTicket for and m.Parking applications.

    a. m.Ticket: To buy public transport tickets and monitor movement of buses and the planning of trips &

    b. m.Parking: To pay for parking on a smart phone.

iv. Mobile platform "Trafi": To plan journeys & select parking locations

Smart energy: Sustainable energy is an important theme of Smart cities. Using clean energy, green energy like renewable energy and conserving energy through efficient use of sensors, analytical tools and actuators controlling energy efficient devices enable to lower carbon emissions and save environment. High usage of energy from biofuel and biomass & high efficiency LED street lighting enable the Smart cities to have environmentally friendly and efficient energy management.

Smart energy management system:

a. Uses sensors, advanced meters, renewable energy sources, digital controls and analytic tools to automate, monitor and optimize energy distribution and usage.

b. Optimizes grid operation and usage by balancing the needs of the different stakeholders involved (consumers, producers, and providers).

c. Involves infrastructure, such as distributed renewable generation, microgrids, smart grid technologies, energy storage, automated demand response, virtual power plants etc.

d. Encourages energy efficient electric vehicles and smart appliances.

e. Employs extended network of intelligent energy devices across a city, with a detailed view of patterns of energy consumption, enabling community-based energy monitoring programmes and improving the energy efficiency of buildings.

f. Comprises of smart grids defined as "electricity delivery system" from point of generation to point of consumption integrated with ICT for enhanced grid operations, customer services and environmental benefits."

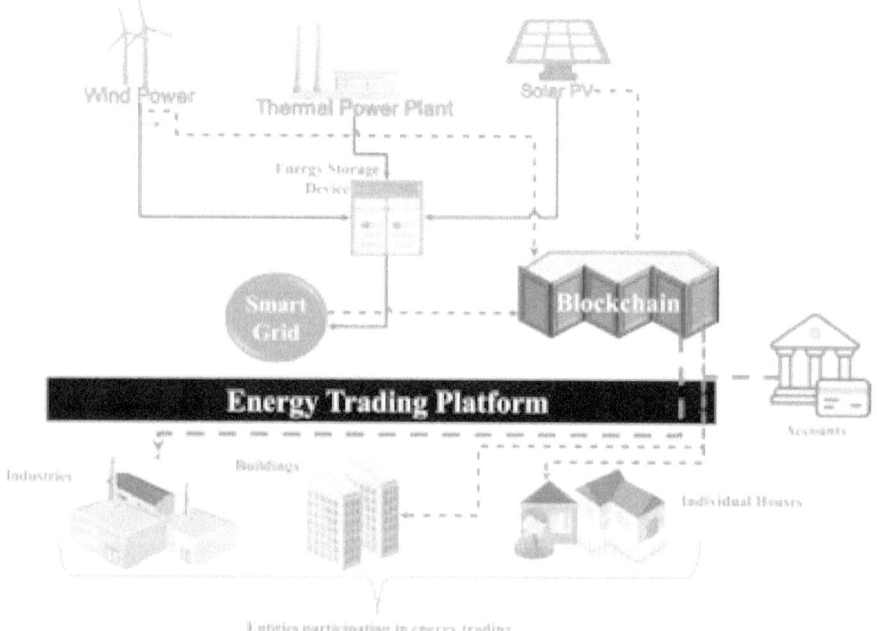

Fig 3.5 Blockchain based energy trading system Fig Source: IEEE Access: Machine Learning adoption in Blockchain based Smart Applications

g. Take part in trading mechanisms for excess generated energy through renewable sources using Blockchain and analytics.

Smart water: In future wars are expected to be fought over the sources of water as the world's population grows dramatically while the fresh water sources are not keeping pace and are rather dwindling in their follows. Smart Water Management involves:

a. Leveraging innovative technologies for better management of water to improve metering and flow management that provides real-time information on the water situation to stakeholders helping conserve water, leading to lower water bills.

b. Using digital technology like remote controlled smart water meters to help save water, reduce costs and increase the reliability and transparency of water distribution.

c. Mirroring physical pipe networks with data and information connection & collection network to analyse flow and pressure data to determine anomalies (such as leaks) in real time to better manage water flow.

d. Encouraging rainwater harvesting, waste water treatment & recycling, reducing evaporation from water sources and increasing potable water through desalination etc.

Smart waste management: Waste reduction, segregation at source, treatment, and recycling are the most important activities for Sustainable development. Hence Smart Waste Management gained a lot of importance. Smart Waste Management facilitates:

a. Efficient monitoring, collection, transport, processing, recycling, and disposal of waste.

b. Facilitates conversion of waste into a resource and create closed-loop economies.

c. Uses integrated sensors in the dustbins and CCTV cameras and enables predictions for routing trucks to the waste

d. Enables the movement of different kinds of waste to be monitored,

e. Leverages technology to better manage the flow of waste from source to disposal.

DWMS (Decentralized Waste Management System) is employed in Smart cities with the following elements:

a. Connected & tracked GIS-enabled refuse collection vehicles.

b. Continuous monitoring and evaluation (M&E) framework through connected staff & public with infrastructure for centralised monitoring systems.

c. Fool proof system for segregation of waste at source from every user.

d. Waste treatment, recycling and composting at all levels from individual apartments to waste treatment plant with video surveillance.p

Smart environment: By leveraging efficient connectivity solutions like very high-speed public Wi-Fi internet connections, the Smart cities strive to provide high quality of life.

a. IoT & data analytics are used extensively in lighting, traffic, air pollution, agriculture, healthcare, retail and logistic for low levels of pollution, clean water and fresh air.

b. Regulatory Sandboxes supported by government enable start-ups to test their products in a low risk environment by experimenting in the captive eco-system comprising of commercial, residential and multi-purpose buildings housed in the Smart city. Successful products can be scaled up for further development and commercialisation.

Smart health: Healthcare is a vital aspect of citizen wellbeing and Technology can dramatically improve the effectiveness of its delivery as follows:

a. Big data can help to develop predictions or identify hotspots of population health in times of pandemics or health impacts during extreme weather events.

b. Analytics converts health-related data from digital health records, home health services and remote diagnosis, treatment, and patient monitoring systems into clinical and business insights.

c. Monitor the health conditions of citizens with the help of patient connected electronic wrist bands by. Remote collection of patient health vitals and data facilitates quick diagnostics and establish automated alert system for patients about medications and health check-ups.

d. Telemedicine and smart phones can combine to take quality healthcare to remote locations and thus overcome shortage of doctors and nurses (see http://www.medicmobile.org).

Smart Citizen engagement: Citizens access all government services through a single easy to use interface. All citizens are encouraged to learn digital skills irrespective of age group and consume government services actively and provide feedback for improvement. The following are some of the vehicles employed.

a. Smart management electronic platforms for communication with city administration engage citizens and business in decision-making.

b. Mobile apps for feedback: Platform for residents to report overflowing bins or objects blocking public roads etc., for speedy remediation of citizen problems.

c. Mobile application to foster the direct communication between the residents and the municipality.

Smart digital layers: The digital layers encompass some digital layers that facilitate data collection through connected solutions. They are:

a. Urban: The layer where physical and digital infrastructures meet. Examples include smart buildings, smart mobility, smart grids (for utilities such as water, electricity and gas) and smart waste management systems.

b. Sensor: This layer includes smart devices that measure and monitor different parameters of the city and its environment.

c. Connectivity: This layer involves the transport of data and information from the sensor level to storage and to data aggregators for further analysis through a robust, reliable and affordable broadband network. For example, LoRa wireless technology supports low data rate communications over long distances by sensors and actuators for M2M and Internet of Things (IoT) applications.

d. Data analytics: This layer involves the analysis of data collected by different smart infrastructure systems, to help predict some events (such as traffic congestion).

e. Automation: The digital enabling interface layer that enables automation and scalability for many devices across multiple domains and verticals.

f. Data management: Provide free access to financial, public procurement, real estate, transport and other open data.

The integration of the various IoT and communication networks achieved through ICCC enables the deliveries of these services through a centralised platform that leverages the efficient ecosystem for the Internet of Things and the capacity to make use of the big data generated. This is discussed in the next section.

Command and Control layer of ICCC: The nerve centre of the ICCC that manages the following:

a. Communication with Stakeholders

b. Device Control (asset, access and authorization)

c. User Interface and Visualization

d. Complex Real Time Event Handling

e. Service Management enabling the coordination of various activities as follows:

Important activities of the Command and Control Centres

1. Traffic & Vehicle Monitoring:

- Track road transport defaulters
- Vehicle Thefts
- Wrong usage of Vehicles
- Criminals and thieves etc.

2. Electricity & Water

- Automated Meter Reading

- Monitor leakages, pilferages, over-use

- Control on and off public utilities as per need

3. Disaster Management:

- Sensitise people in times of distress with appropriate alerts

- Coordinate relief measures by Govt. machinery

- Direct fire brigade, ambulance, and police to the right spot speedily

4. Safety & Security

- Deter criminals proactively

- Ensure proper safety for women & elders at night.

- Track & capture criminals

Singapore being a Digital Nation and a Smart city stands as the best and glorious example in the world to lead the way and be a role model to follow. We shall examine the case of Singapore in the coming sections of this book. The focus will be on select areas from the earlier mentioned figure, which are critical for Sustainability and in which Singapore offers a role model to the rest of the world.

Smart City Governments and Best Practices Across the Globe

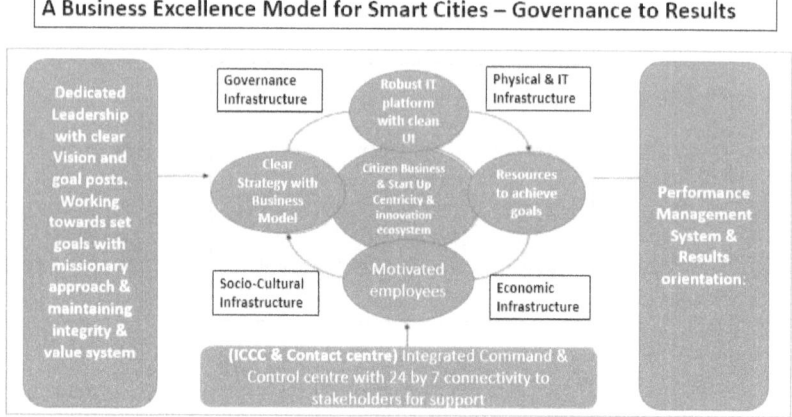

Fig 3.6 Business Excellence model for Smart Cities (Leadership to Goals achievement)

The different aspects of the Business excellence model components that integrate with each other that translate the vision and goals of the leadership are mentioned below.

Vision: A clear road map for the future with a desired state and a time frame to achieve the same.

The following approach is undertaken by the Governments across the world for achieving the mission of Sustainable & Vibrant Smart cities:

1. **Define:** Smart city definitions can vary across the world depending on the community traits and local issues. Hence it is important to clearly define the vision and the goals that are best suited to the ground situation as this may motivate the following action plans and allocation of resources. Things like geography, definition of City/Community, links between cities and countryside and flows of people between them could be different in different situation and hence they need to be considered for the definition.

2. **Understand in detail:** This is like the 'Measure & Analyse' step in DMADV approach in Six Sigma. Various demographic, socio-

economic, cultural and educational levels of the population have an impact on the 'Motivation' behind and 'Expectations' from a Smart City initiative. Hence these should be laid out with absolute clarity before framing the policies. Study the community to know the citizens, the businesses' needs, know the citizens and the community's unique attributes, such as the age of the citizens, their education, hobbies, and attractions of the city. A thorough SWOT analysis of the city can also lead to a tailor-made vision for the right impact.

3. **Develop a smart city Policy**: Develop a policy to drive the initiatives, where roles, responsibilities, objective, and goals, can be defined. Create plans and strategies on how the goals will be achieved.

4. **Engage**: Validation of your policy is achieved when Citizens are happy. Continuously involve the subjects, implement activities for impact, vary as per the feedback of the Citizens: This can be done by engaging the citizens through the use of e-government initiatives, open data, sport events, feedback surveys, decision impacting involvement through transparent access to & voting on Government activities etc.

Sustainable development is one of the most laudable objectives and outcomes of the Smart City initiatives. Singapore sets an example to the world in implementing Sustainable development strategies with a vision of a 'Vibrant, Sustainable and Thriving economy for the next 50 years' an excellent vision to have. In the subsequent chapters we will explore how Singapore is progressing towards this vision through a multipronged approach of Water management & Environmental Pollution management through activities like,

a. Treatment & reuse of reclaimed water, rainwater catchment systems, and desalination,

b. Minimize pollution and crowding, through "meaningful transportation" principle that encourages citizens to reduce Oil guzzling personal transportation in favor of hared multimodal transport.

Leadership: Focused and dedicated management that leads the Smart city projects with a clear operational accountability, fair policy management framework, high standards of Governance with competence & continuity is very much required. The leadership should work with a democratic & subject (citizen, start-up and business) centric approach & highest levels of accountability & responsibility while maintaining fairness to achieve the vision laid out.

Singapore's Centralised leadership approach executed through the Singapore's Smart Nation Office under the aegis of the Prime Minister's Office enables better coordination across all government ministries resulting in a coherent & effective approach. While this approach has many advantages that simplifies communication, increases efficient execution through accountability, centralizing and clarifying data ownership, a decentralized approach could be more effective in large cities like New York & London. Through effective interagency collaboration across different empowered, independent & specialized departments, New York & London have demonstrated excellent progress in implementing complex Smart city solutions and solve difficult problems that arise with time, growth & maturity.

Resources: Clear business model with adequate funding for the projects & various activities through government & public – private partnership sources in a sustainable manner provides necessary resources for the Smart City's development.

Large private companies that are stakeholders in a city's development like big employers in the vicinity, with a large potential customer base, Providers of IT solutions, hardware, software and middleware for Smart city development, particularly those in the technology space, are keen to participate in the area development activities. They can offer a variety of specialized resources & invest time, money, solution development, offer space and people to implement activities & conduct research aimed at helping the smart cities they are in.

Activities like hackathons and competitions act as merit-based funding sources for start-ups and citizens while developing valuable solutions that attract employment opportunities and funding to their cities.

Strategy: A coordinated and integrated approach to garner public & private sector involvement to work intensely & with involvement towards achieving the goals laid out while managing the continuity of resources.

A clear strategy includes an integrated approach through a variety of programs implemented with short term and long-term objectives in mind to involve citizens, develop employees and improve capabilities.

To develop right strategies, cities undertake massive collaboration activities with citizens, thought leaders, leading organisations, and other cities. Some of them are listed below.

a. Host smart city conferences and expos that bring together organisations and thought leaders for cutting across silos and evolving practical solutions,

b. Participate actively in study trips and round tables to share their learnings with other cities,

c. Curate international good practices and benchmark with the best,

d. Undertaking joint ventures & collaborative planning with organisations,

e. Involve experts with specialized knowledge and work on developing industry clusters that act as magnet for investment, talent & employment.

Helsinki's 'The Six City Strategy,' together with Espoo, Vantaa, Tampere, Oulu and Turku and its extensive collaboration with other European cities through Horizon 2020 projects, EIT Digital and the EIT Climate-KIC & Smart City expos stand out as an example of collaborative efforts.

Employees: Motivated and competent employees who are always well incentivised to provide the best services. There should be accountability at all levels for all projects and activities.

It is very important to develop capabilities among citizens and employees to develop, manage and achieve the Smart Cities vision. Continuous Learning and Development is an important aspect of employee development. The following are some of the programs that focus on developing Smart City specific capabilities among the potential employees:

i. Vienna's – Smart City Master's degree for development and implementation of tech-enabled integrative urban solutions.

ii. The Singapore Management University's Smart City Management and Technology Major.

iii. European Universities' Joint Master's program in Energy for Smart Cities actively engages start-ups and industrial partners to develop practical & implementable solutions.

Citizen services: Focus on better customer services, improve quality of life, single window connectivity platform, training and cultural development, health care, education, Security etc.

• London and Singapore undertake extensive citizen engagement program to spread computer literacy across all segments of their population for digital inclusion. This will enable them to access and consume online government services with ease.

• Hamburg's Smart Citizen CoLab & Melbourne have engaged Citizens to discover new services & develop smart city applications and projects through crowd sourcing of ideas. Continuous engagement of citizens to co-develop solutions and offering them access to the city's management process empower the citizens leading to all round development & happiness.

• Boston's playbook, envisages leveraging 'smart technologies.' help governments become friendly to citizens by becoming "more beautiful, more delightful, more emotionally resonant, more thoughtful, and more pleasurable to interact with—not just cheaper."

Business services: Provide a good atmosphere with ease of doing business and encouragement to invest and create employment and value addition.

Transparency offered through availability of large sets of data that enables them to plan activities, know status of project approvals and the potential opportunities is a big boon to the businesses. Open data shared in multiple formats powered by different Business intelligence tools offered by the likes of 'Transparency portal' of Helsinki Smart city allow users to filter and cross-check a range of variables. The portal offers over 200 heat maps, historical maps, and 3D maps of interlinked data that can be easily understood and manipulated.

Innovation & Start-ups – Vibrant atmosphere for encouraging start-ups, innovation, and ideas. Ideas should be encouraged, collected, analysed, implemented and rewarded in a sustained manner.

Singapore developed innovation districts & centers like the Punggol Digital to encourage innovative solutions & generate employment as well Regulatory Sandboxes with the involvement of the government offer a safe testbed to manage risks and develop new solutions

Seattle partnered with the largest University in its area, The University of Washington to set up an innovation zone to develop and design Citizen centric Smart technology solutions

Processes & Systems: Stable and robust automated IT platform to coordinate, deliver, collect data, analyse information, and facilitate better delivery of services

Helsinki set up an open data platform with a variety of visualizations & tools to measure and present its performance to citizens, startups, and businesses drive new innovations. This has also put a check on partisan policies thus improving transparency and accountability in Public procurement activities saving a large amount of money.

Performance Management System & Results orientation: Clear laid out **SMART goals** against which performance is continuously measured and corrective action taken, facilitated through a robust IT system offering a detailed view of the status through Private and Public Dash boards.

Montreal Smart city has an "Open by default" data policy that offers utmost transparency to its stakeholders and has developed new tools to make its key performance indicators for its budget, project monitoring, and service level targets, available to the public.

It is imperative that the Smart cities create a comprehensive Business excellence program that also compares their progress across its various dimensions on a periodic basis while at different levels like State and Country, there could be a competition amongst the Smart Cities or areas within the Smart cities over a subset of parameters and prizes and rewards are announced.

The Smart cities function to achieve growth objectives through a holistic development of different types of infrastructure, which together offer the best quality of life to their citizens. These different types of infrastructure namely Physical Infrastructure, Institutional and Governance Infrastructure, Socio-Cultural infrastructure & Economic infrastructure which need to be developed together are represented in the following diagram.

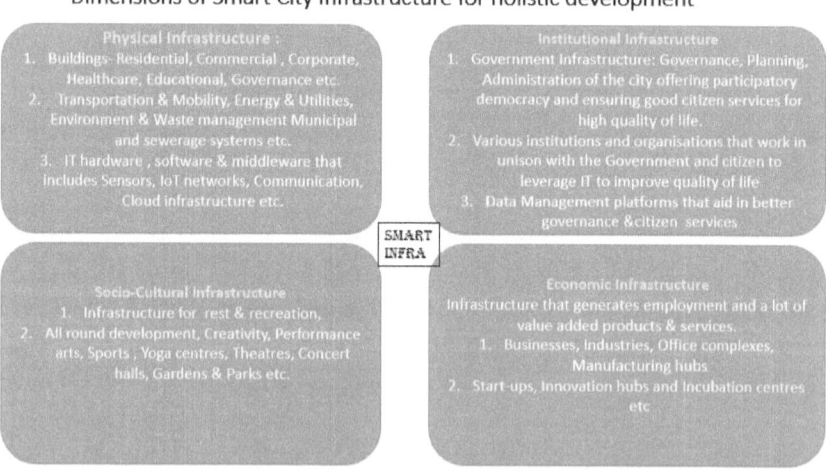

Dimensions of Smart City Infrastructure for holistic development

Fig 3.7 Dimensions of Smart city infrastructure for holistic development

Model Smart Cities Across the World

- **Singapore:** Smart city activities are led under the guidance of the PM's office through SNDG, the Smart Nation and Digital Government Office ensuring highest level of attention, thus enabling for strong coordination among the various activities to work synergistically, aligning with the overall vision of the program.

  Substantial budgets amounting to an average of 500 million SGD per year are devoted for investing in a strategic manner to reach the Smart City objectives.

  Several pioneering programs targeted at different stakeholder with far reaching objectives aimed at an all-round development of the Smart city like:

  i. Productivity Solution Grant to fund 70 percent of the costs of adopting technology to SMEs in their work processes,

  ii. Digital Skills program to facilitate improving IT skills in all age groups through a variety of activities,

  iii. Tax incentives for R & D and IP development investments,

  iv. Tech skills accelerator and Advanced Technology internship programs to develop the capabilities in the emerging technologies,

  v. Innovation districts and Regulatory Sandboxes with Government involvement and investment to develop high potential ideas as scalable solutions to real life problems etc.

The above are a sample of many programs aimed at citizen involvement, provision of single window services with easy access, encourage innovation & entrepreneurship, that have enabled Singapore to be the best among Smart Cities across the world.

- **London:** Smart City initiatives are handled through accountable & dedicated entities with delineated responsibilities assisting the Mayor like:

  i. The Smart London Board – Industry experts and thought leaders to envision, strategize & implement activities for reaching the goals of Smart City development.

  ii. London Office for Technology & Innovation (LOTI) – To study global best practices for benchmarking and integrate the learnings into the Smart city coordination & development for being on the forefront of Smart City development.

  iii. Chief Digital Officer: Leads digital transformation activities of the city like providing secured & free Wi-Fi to all citizens, undertaking digital literacy activities & inculcate digital skill in citizens of all age groups to enable them access the e-governance portals and services through them, and improving their participation in co-creating the projects for continuous improvement.

The accountability, decentralization, focused approach, investment of resources strategically and huge citizen involvement have resulted in a superb impact in making London, a leading Smart City in the world.

- **Seoul:** Part of one of the most Technologically advanced nations, South Korea that houses global leaders in Electronics and Automobiles, Seoul Smart city leverages all its strengths to offer high quality seamless services to its citizens. Some if its programs are outlined below.

  i. "Global Digital Seoul 2020: Smart City Seoul with New Connectivity, New Experience" plan for ensuring a highly involved & evolved Citizen led Governance.

  ii. Mobile Seoul website: Single interface to access, consume and experience multiple services in one place and lists over 60 real-time services in 11 categories, including mobility & transport, bus and subway operations, social & cultural events, employment opportunities, real estate information, and many other public services.

iii. FixMyStreet: Registering geotagged citizen complaints and resolving them.

iv. Oasis: Encourage citizen led innovation by registering their ideas for improving their city with a dedicated support system to turn them into reality.

Many such innovative approaches have led Seoul to become one of the most acclaimed smartest cities in the work in line with its potential.

- **New York:** World's best financial powerhouse and the most valuable city, New York has excelled in its Smart city management practices with novel approaches as outlined below.

  i. Decentralized approach: Collaboration and interdependence among multiple empowered departments to work with a common vision of achieving smart city goals.

  ii. OneNYC: Unified city plan aligning and working in a coordinated manner to achieve multiple goals for a sustained city development like diversity and inclusivity, equity, growth, resiliency, and sustainability.

  iii. NYCx, which sources tech-knowledgeable individuals to share experience and expertise when applying frontier technologies to smart city initiatives.

- **Helsinki:** Leveraging open data approach and collaborating with many like-minded cities and projects, Helsinki can be considered the most transparent Smart city in the world. Some of its notable initiatives are outlined in the following.

  i. Forum Virium Helsinki: Innovation division of the city FVC plans and implements open government and transparent policies for achieving all around development in line with Smart city goals.

  ii. Transparency of public information: Extremely high levels of interaction between the Smart city management and the citizens across the city through a variety of channels like face to face meetings & webinars involving all City Council meetings that impact them offer highest levels of transparency. Further the citizens are free to interact with the leadership and offer feedback/contest on the decisions in real time.

iii. The Six City Strategy open innovation platform: Collaborative approach to pool resources & ideas with other global leading Smart Cities to come out with new solutions and better approaches for benefiting their citizens.

iv. Europe wide collaboration through multiple programs: Horizon 2020 projects, EIT Digital, and the EIT climate-kic. Kalasatama's aim is to save one hour of every citizen's daily time.

v. Collaborative Innovation: Value of Open data is unlocked in multiple ways through an active collaboration between Helsinki's departments, residents, citizen organizations, industry, SMEs, start-ups, to come out with innovative solutions that offer smart and clean services. This has led to a large number of projects that can be implemented across the world in the domains like infrastructure, buildings, and citizen well-being.

vi. Citizen Centric Mission oriented Projects: Nifty Neighbor, Helsinki Well-being, Smart & Clean Helsinki Metropolitan and many more such projects are conceptualized and implemented for enhanced quality of life for all citizens.

- **Amsterdam:** Tremendous collaboration with the citizens has enabled Amsterdam to offer phenomenal services to its citizens and save a lot of money. Some of the best practices are as follows:

  i. Collaboration with citizens: Over 170 projects collaboratively developed by residents, government and businesses running on an interconnected platform through wireless devices facilitating real time decision making.

ii. Amsterdam Smart City Challenge: Multiple solutions developed through competitions and hackathons and run on the platform to reduce traffic, save energy & improve public safety.

iii. Innovative Mobile applications: Mobypark, a citizen centric mobile park facilitates smooth parking and enables estimation of traffic density.

Yokohama Smart city Project (YSCP): YSCP is set up by Japan Government through its entity, New Energy Promotion Council in collaboration with leading companies in the world as model smart city project in one of its largest townships with 3.7 million citizens, Yokohama. YSCP is setting standards in environmental friendliness through its focus on Smart Grid (Self Managing Reliable Transmission grid) technology, a model of managing electricity consumption in Smart Cities to achieve sustainable energy consumption technology while reducing carbon footprint.

The following are the salient features of the project:

1. **Partnerships:** Collaboration with leading companies like Accenture, Toshiba, Tokyo Gas, Nissan Motor, Meidensha, Panasonic etc.

2. **Crash Carbon Emissions:** Focus on reducing emissions by over 30% through focus on non-polluting electric vehicles.

3. **Renewable energy focus:** Reducing dependence on Conventional energy and focusing on Sustainable, environment friendly Renewable energy.

4. **Eco-friendly holistic concept, WAEM (Wide Area Energy Management):** This involves co-ordinated approach to balance energy consumption by combining Home, Building and Community energy management systems.

   • **HEMS** – Home energy management system offering transparent household consumption with real time visibility.

- **BEMS/FEMS** – Integrated Building or Factory management system to synergise the working of co-generators, storage batteries, and electric vehicle charging and discharging infrastructure in large apartments, factories, offices and commercial complexes to optimize energy consumption and

- **CEMS** – Integrated energy management of a township combining several households and buildings & housing several new generation energy conservation technologies like Electric vehicle charging stations, supervisory control, and data acquisition (SCADA) systems, electric storage batteries, and undertake large scale solar energy generation.

Through this experiment, YSCP set an example of employing energy conservation practices for a sustainable environment with minimum carbon emissions, proving the opportunity ahead for Smart Grids.

San Francisco: Being closest smart city to the world's leading innovation hub, Silicon Valley in USA, one can expect to see the best of smart city applications in action in this city. San Francisco Smart city management was the finalist in the US Department of Transportation Smart city challenge and was awarded a grant of 10Million US Dollars to focus on technology innovations that make mobility smarter & equitable. It has identified transportation and mobility as a focal area to impact citizens' lives, impact environment, reduce costs and save money through the following steps:

i. Encourage shared transport over owner driven.

ii. Reduce transportation emissions through use of electric vehicles and reducing traffic congestion.

iii. Reduce collisions and fatalities through better traffic management.

iv. Reduce cost of transportation as a share of domestic cost.

Some of the interesting applications developed by San Francisco's Smart city team are:

a. SFPark, a large-scale controlled parking pricing experiment to optimally utilise cities parking spaces.

b. Sigfox to deploy a new generation of IoT communications network wireless platform, engaging entrepreneurs, and catalysing start-ups to experiment with long-range, low-power, inexpensive connectivity.

c. IDEO workshop involving intrapreneurs to develop solutions for earthquake preparedness and emergency response.

d. LightHouse for the Blind and Visually Impaired to enable the visually impaired to use the airport services without a problem.

San Francisco can be looked upon for cutting edge solutions that can be emulated by the rest of the world to improve their citizens' lives.

- **Montreal:** Believing in utmost transparency and democratic processes, this city in Canada implemented transparent and accessible data sharing activities and collaborative public spaces. Montreal implemented several 'Participatory Democracy' projects under its Smart City Action Plan. Making Open data while protecting the Privacy and security of personal data has enabled citizens to track its performance and develop immense trust in their government.

- **Boston:** The research lab for Artificial Intelligence projects across the world and the world's most coveted city for top class University education, Boston is leading the research in developing innovative cutting edge technology solutions to solve peoples' problems leveraging artificial intelligence, autonomous vehicles, and robotics.

- **Melbourne:** Developing innovative solutions to help disabled citizens feel comfortable is the hallmark of Melbourne's Smart city activities that collaborates intensively with its citizen through its Town Hall project. It integrated the city's open data with smart assistants like Google Assist and Amazon Alexa, to provide updated information through voice, text, as well as screen readers, to enable disabled feel normal.

- **Barcelona:** Barcelona the most connected city in the world invested considerably in IoT applications for the city. It is building unified operating system that would run the entire city on a single interface. Barcelona is leveraging IOT extensively and is realizing the benefits of its investments in cutting the costs and saving huge energy through automation. A variety of services are developed by leveraging the data collected from the sensors and connected systems that substantially improve convenience, quality of life, cutting down costs, wastages and provide high levels of Return on Investment. Its LED project has saved over 30% of energy costs. The WHO has studied Barcelona intensely and concluded that "IoT-powered smart cities stand better chances of becoming healthier cities. Thus, Barcelona has proved the world that Smart City initiatives can be economically sustainable and viable and serve as a model for other cities in technology-led urban transformation and city management through its ROI on IOT investments.

Spain is also a pioneer in implementing Automated Waste management systems. It has implemented the practice of 'Reduce, Reuse and Recycle of waste and facilitated the existence of cleanest cities in the world.

An automated vacuum waste collection system, also known as pneumatic refuse collection, or automated vacuum collection (AVAC), transports waste at high speed through underground pneumatic tubes to a collection station where it is compacted and sealed in containers. When the container is full, it is transported away and emptied. The system helps facilitate separation and recycling of waste.

Pneumatic refuse collection in
Vitoria-Gasteiz, Basque Country,
Northern Spain

Fig 3.8 Automated Vacuum waste collection bins (Source: Wikipedia)

The process involves Post Box like bins that collect the segregated waste from the citizens and through an underground suction system the waste travels to an automated central processing facility. There, it is directed to landfills or composting plant or waste to energy conversion plant after final segregation.

- **Shanghai:** Excellence in Artificial Intelligence, is the hall mark of Shanghai that operates as a lab for developing applications using most advanced technologies to monitor, track and benefit the citizens through ultra-high levels of security and high-quality medical services. With 100 percent city-wide optical fiber coverage, it offers 200 public services on cloud and piloted data-sharing across different government departments. All government services, regulatory approvals and registrations are offered online with zero downtime. Its Integrated Smart City Management control center implements a lot of state-of-the-art citizen services, some of which are outlined as follows:

i. Analytics and Integrated health record management that stores billions of electronic clinical records of its citizens enables to offer cost effective and pin-pointed treatments. Public healthcare institutions in Shanghai share their databases to form the largest personal health data center in China, storing over 30 billion clinical records. Health data analytics is also playing a major role in adjusting healthcare prices and in assessing quality of hospitals.

ii. Traffic visualization through heat maps enables better traffic redistribution and minimize congestions and improves fuel economy.

iii. Big data analytics enables the government to put a check on traffic mishaps and offences.

Summary

This chapter covered Smart city aspect in detail and examined several facets of the Smart city like:

1. Definition of Smart City and its place in the modern world's evolution.

2. Important components of Smart City and their utility.

3. Layers of Command and Control Centres and their role in a Smart City.

4. Different infrastructure elements of Smart City.

5. Different Smart services offered by a Smart City to its citizens.

6. Smart City Business excellence model.

7. How to set the mission and goals of a Smart city.

8. Smart city best practices across different elements of the BE model.

9. Best Smart cities in the world and what differentiates them.

CHAPTER 4

How Blockchain can Secure IoT Powered Smart Cities

Introduction

Internet of Things (IOT) is deemed to be one of the fastest growing phenomena in the coming days with estimates pointing to tens of billions of connections being added to the internet through connected things that define the way we live in the future.

IOT is deemed to have widespread usage in industries, manufacturing, retail, consumer, health care, smart city, transportation & logistics, energy management, agricultural, environmental, financial, automobile, government and military applications.

As the role of IoT in our day to day life grows, the data that is being generated by the humongous number of these connections whether it is in the form of sensors or autonomous vehicles or robots or health bands is growing by leaps and bounds. Mining of this data can yield a huge value, thus making these dumb devices meaty targets for cyber criminals. Though the advent of 5G technologies & increased adoption of Edge computing & Fog computing are aimed at providing a high amount of processing power at the device level, there exists a tremendous risk that these networks of devices could be highjacked as has been demonstrated in several cases. In this chapter, we shall examine the risks of IoT powered systems that can also put into jeopardy, our lives in the future Smart cities and how Blockchain can help in balancing this risk of Cyberattacks.

Security Breaches on IoT Powered Systems

IOT also becomes an easier target to the malware attackers to penetrate and flood the internet with unwanted traffic creating nuisance to global population at large. The havoc caused by the Mirai virus that attacked a mere 6 lakh devices to shut down many publisher sites in October 2016 is a pointer of the things to come.

The following figure gives a brief overview of the Mirai attack that shut down several websites and portals globally in September-October 2016:

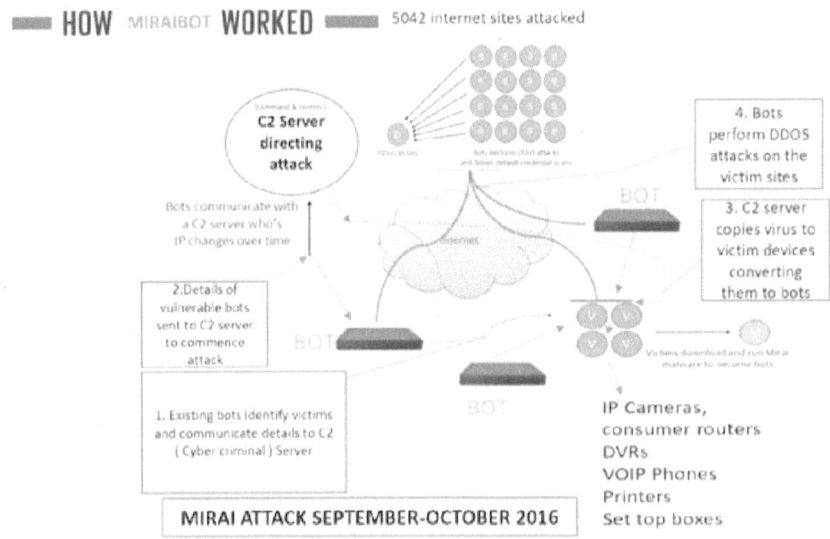

Fig 4.1 Mirai Botnet attack exploiting IOT Vulnerability for DDOS attack

There have been several instances, where, cyber-attacks have been used as a weapon of state sponsored act to penetrate adversary computer systems and shut down their activities.

StuxNet was one such attack apparently conducted by US & Israel forces on Iranian Nuclear weapon program.

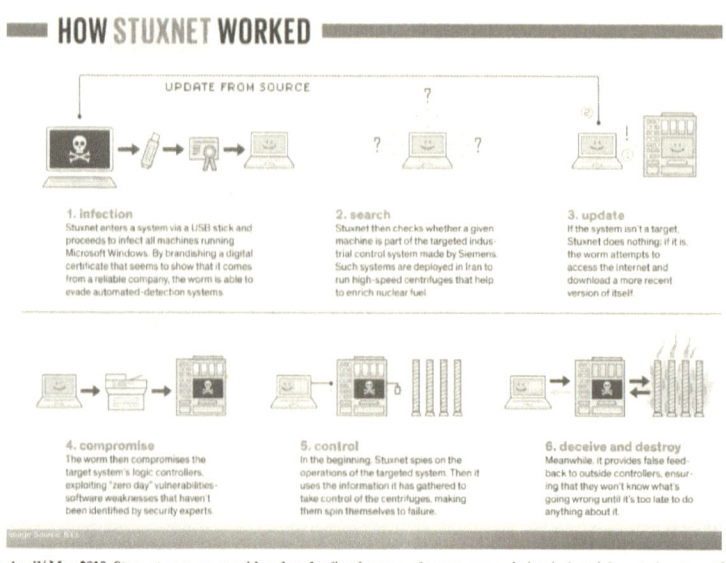

In April/ May 2010, Stuxnet a worm, considered as the first known cyberweapon, to infect industrial control systems (ICS) reportedly developed by US & Israel, targeted programmable logic controllers (PLCs) that control the automation of electromechanical processes, such as those used for centrifuges infected & destroyed over 1000 Iranian Uranium centrifuges

Source: https://www.cyber.nj.gov/threat-profiles/ics-malware-variants/stuxnet

Fig 4.2 STUXNET Worm Penetrating enemy defences to neutralise weapons

The advent of the 'Smart City' concept is leading to a number of IoT devices being used across the landscape of modern infrastructure for leveraging automation of building, home, corporate and convenience facilities. The vulnerability of many of these IoT devices has been well exposed in a research study undertaken, namely the 'Chain Reaction' attack.

Zigbee Chain Reaction Attack

Chain reaction is an academic simulation project that demonstrates the vulnerability of a set of IoT devices to penetration by unauthorized worms, that can enter the system through a single device like a Philips Hue bulb. By gaining an unauthorized access through a single infected bulb and placing it in a network controlled by ZigBee protocol, the hackers were able to take over an entire IoT network of devices showing the vulnerability of an entire Smarty populated with different IOT devices. Though the vulnerability associated with this device was later plugged, it demonstrates the real cyber threat faced by IoT dominated world in future where a new types of vulnerable communication networks are created in addition to the numerous already existing networks. (**Source:** https://eprint.iacr.org/2016/1047.pdf).

Hence it is expected that Blockchain adoption is a must to protect the IOT devices. Needless to say, the growth of Blockchain technology will therefore closely track the growth of IOT devices.

Some of the interesting applications where Blockchain is being leveraged to protect the IOT devices are outlined here.

a. **ENERGY:** Blockchain can be used as an immutable and trusted repository of data for the source and performance of various renewable energy devices. Smart contracts can be used for trading and exchange of energy produced while, cryptocurrency could be used as a medium for facilitating the trade between the producers and suppliers.

b. **BUILDING AUTOMATION:** Blockchain can be used to identify and offer trusted access to IoT devices that are used extensively in facility automation projects and facilitate authorized interactions with the devices for building access controls, parking lot automation, energy & utility management.

c. **INDUSTRIAL AUTOMATION:** Blockchain can serve extensively in 3D manufacturing by serving as a trusted repository of patented & specialized designs, facilitating trusted interactions between machines and in tracking the provenance of genuine spare parts. Robots and drones can be managed in a secure manner to deliver desired outcomes without the threat of being hacked for unwanted outcomes.

d. **MOBILITY:** Blockchain can facilitate tracking of vehicles on the move, vehicles could be identified, and various fees collected by authorities towards, toll, parking or traffic violations as well, without any room for dispute and lack of trust.

e. **DISASTER MANAGEMENT:** Drones can be used extensively for various purposed like photography, environmental hazard tracking, security and mapping in geographic information systems. Blockchain can be used extensively in ensuring authorized usage of drones. By mapping the information and visual records collected to various authorities like police, emergency services, fire department, local self-governmental agencies etc., through a Blockchain enabled distribution ledger, incident responses could be done in a rapid manner to save a lot of time and effort in times of distress.

f. **INSURANCE:** Blockchain can be extensively used to track and monitor the insurance record management and redressals of cases in all industries and for all automobiles.

g. **CONNECTED CARS:** Bosch has developed Blockchain applications to manage automobiles through smart phone applications to safely and securely track and record various performance and status related features of the automobile like odometer readings, fuel tank readings, mileage of car and tyre performance etc., undertake preventive maintenance and prevent odometer frauds. associated with drivers maintaining fuel usage records, travels and used car transactions.

Blockchain Based Security for IoT Powered Systems

In the following section, we shall examine how Blockchain can protect the IoT systems, through a variety of applications, use cases and architectures.

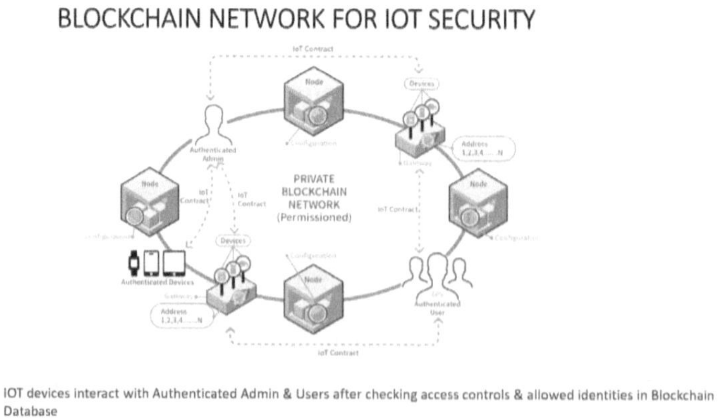
BLOCKCHAIN NETWORK FOR IOT SECURITY
PRIVATE BLOCKCHAIN NETWORK (Permissioned)
IOT devices interact with Authenticated Admin & Users after checking access controls & allowed identities in Blockchain Database

Fig 4.3 Authentication and Authorisation through Blockchain to mitigate IoT vulnerabilities

IBM Watson IoT platform is an excellent initiative by IBM to seamlessly integrate the IOT devices to the Blockchain eco-system to enable them to enjoy the benefits and safety of distributed ledger applications. By using Watson IoT Platform on Blockchain, Internet of Things (IoT) devices can send data to and invoke smart contract transactions on IBM Blockchain Platform or on the open source Hyperledger Blockchain.

Securing Robots Through Blockchain

Singularity Net, led by Dr. Ben Goertzel is founded by Hanson Robotics, one of the world's largest Humanoid robots company. Hanson robotics is the company behind the most popular humanoid robot on earth, namely Sophia, which is even granted a citizenship of Kingdom of Saudi Arabia. The following is an excerpt of an on-stage interviews between, Dr. Ben, Chief Scientist of Hanson Robotics and Sophia:

Dr. Ben: "Sophia, What software are you running?"

Sophia: I am using the Hanson AI software stack that included the Open Cog General Intelligence engine as a component and Singularity Net at the backend.

Dr. Ben: What is Singularity Net?

Sophia: Singularity Net is a Blockchain based platform and a marketplace for AIs. It supports intelligence on the emergent level of the whole network, as well as the level of the individual agents.

This then is, the future of secured automation. The Blockchain based backend system allows access to, from and between robots in a safe and secure manner so that no malware attackers and unauthorized cybercriminals have access to the agents of automation to cause havoc to the mankind.

This also reminds of Technological Singularity, a hypothetical moment in time when any physically conceivable level of technological advancement is attained instantaneously. At this point which many experts predict to happen within the next 20 years, the self-directed computers will develop super intelligence with their intelligence increasing exponentially rather than incrementally. This, in case if it really happens, is expected to transform the life on earth and can also enable them to find solution to many human problems including disease and mortality.

The only way in which this level of intelligence that is already growing exponentially to be secured and controlled is through the risk management and protective powers, the Blockchain technology offers.

The internet of Robots marketplace secured through Blockchain allows for a safe and secured access to ensure that proper verifications and multi-level protection is provided to the Robots to ensure that they are used only for helpful and positive activities that benefit the mankind.

The same can be extended to all autonomous objects for a breakthrough management in a secured manner, as shown in the figure on 'Decentralised Management of Autonomous objects through Blockchain.'

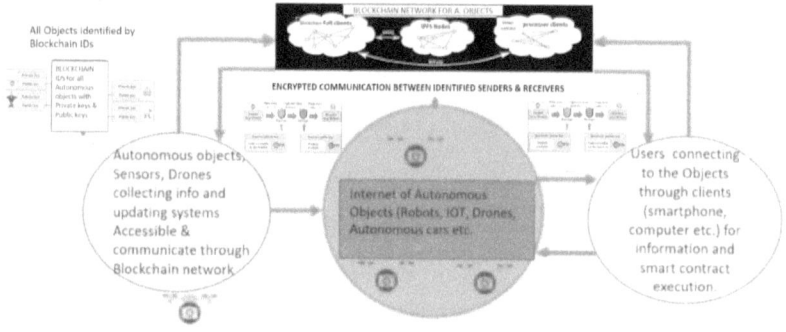

DECENTRALISED MANAGEMENT OF AUTONOMOUS OBJECTS MANAGED THROUGH BLOCKCHAIN

Fig 4.4 Internet of Automation Agents: Communicating through Blockchain

Secured Access and Management of Automobiles Using Blockchain

Autonomous driving vehicles are expected to come onto roads in large numbers and are an amazing extension of IOT technology combined with artificial intelligence and machine learning applications. However, they are extremely vulnerable to hacking by cyber criminals. Blockchain offers the best possible security to the autonomous vehicles to ensure that, they are not manipulated and their owners are not held to ransom.

In December 2017, an exciting disruptive technology company, XAIN and the global leader in automobiles, Porsche announced a partnership to take Blockchain technology to the management of cars.

XAIN and Porsche successfully tested a proof of concept in which an Ethereum client is fused to the car's systems and is connected to the Blockchain network comprising of IPFS and BAAS nodes in the azure market place. The car is tracked and managed through smart contracts and owner wallet present in the smartphone of its owner. The car's systems are tracked and all the parameters recorded in the vehicle wallet that keeps track of various aspects about the car's performance and activities etc.

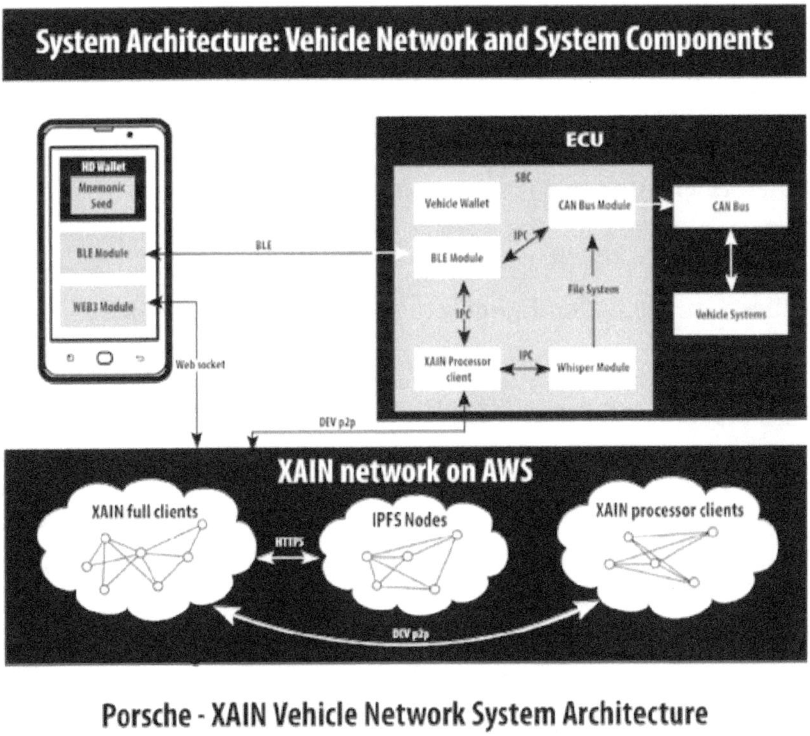

Porsche - XAIN Vehicle Network System Architecture

Fig 4.5 System architecture of Blockchain powered car management implemented by Porsche & XAIN

Figure courtesy: Porsche Digital Lab (https://medium.com/@ porsche_tech)

Source: https://medium.com/next-level-german-engineering/the-porschexain-vehicle-Blockchain-network-a-technical-overview-e1f48c40e73d

The system will allow the authorized owners to access and communicate with their cars using the smartphone connected to the network and do the following from anywhere in the world through internet or through Blockchain powered direct offline connection in a secure manner:

- Lock, unlock doors and luggage compartments from distance securely,

- Communicate with other cars in the network and exchange information,

- Record and manage all critical information on a decentralized trust less system and

- Prevent hacking by cyber criminals.

Another interesting use case is that of managing the autonomous vehicles like drones for the welfare of citizens and protection of these vehicles for unauthorized hacking to use them for illegal & criminal activities. One such illustration is given in the following section.

Securing Drones Through Blockchain

According to a study by Transport Systems Corporation UK, a research conducted by them along with Sheffield University offered a breakthrough solution for controlling Unmanned Aerial Vehicles by using Blockchain. Ability to control Drones, track and record their movements immutably and issue instructions only through secure authentication and access protocols can assist the security authorities in controlling illegal drone activities and ensure that the flight information can be audited & all safety standards can be adhere to. This can also lead to substantial improvement in usage of Drones for a variety of applications including ecommerce deliveries, media and movies etc.

Secured Drone network for disaster relief through blockchain

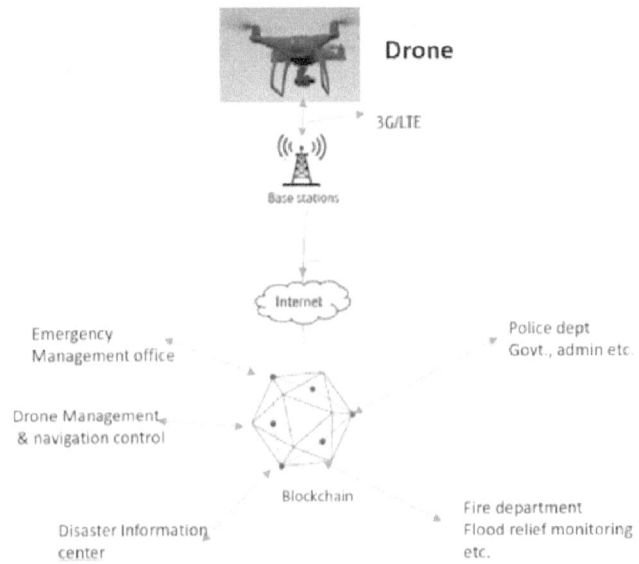

- Access is controlled to drones to prevent misuse of drones for uncivilized and criminal purposes
- Rapid response through shared imagery & related information for disaster management response across government machinery & infrastructure companies
- Shared updated information regarding fire, floods, damaged roads, power outages and bridges etc., helps in quick & timely relief to affected citizens

Fig 4.6 Secured Drone Management using Blockchain

Blockchain Real Life Use Cases in Smart Cities

Distributed ledger based unique Digital identities, Central bank digital currencies and Data exchange frameworks form the key components of Blockchain applications in Smart City projects.

With increased digitisation and huge number of connected dumb IOT devices on the platform, there is a high possibility of malware attacks to hijack the systems and steal the value generated through data breaches, ransomware causing DDOS attacks.

Blockchain with its ability to provide multi-layered security through cryptographic protection for access control, authentication and authorisation while eliminating single points of failure through a decentralised approach, offers a great risk management approach for Smart City applications.

Here we will see a few examples of how a country like China, that has been pioneering solutions in emerging technologies like IOT, Blockchain, AI, ML & Data Analytics is leveraging Blockchain.

Problems in Smart Cities

1. The operations of the public information infrastructure results in the following section. difficulty in coordination among all parties.

2. Data is often replicated and maintained in silos across different departments leading to duplication of efforts.

3. The multitude connected devices & identities interacting across the systems and generating valuable data, pose meaty targets to malicious cyber criminals putting security at risk.

4. Limited transparency exacerbated due to millions of permutations of combinations for connectivity, lead to huge operations costs which can be drastically reduced if information is shared in real time.

Benefits of Blockchain for Smart City Projects

1. Distributed ledger technology overcomes the problems of data centralisation thus, de-risking with respect to ransomware while overcoming limitations due to imperfect information infrastructure.

2. Real time sharing of data across participants is made possible thus unifying the data silos.

3. Immutable and tamper features of blockchain can improve trust, increase transparency & efficiency for a better cooperation & collaboration between departments.

4. Smart contract-based transactions will dramatically improve scalability of transactions at the same time eliminating unwanted human interference. This offers immense flexibility for smart cities and makes them viable.

Some case studies of Blockchain application in Smart cities are given below:

Healthcare in Smart Cities

Benefits of Blockchain for Healthcare:

1. Strengthen the supervision and transparency of the drug supply chain, help trace the origin of drugs and ensure quality of drugs.

2. Strengthen medical data communication, promote data sharing, and protect medical data privacy.

3. Use smart contract technology to formulate medical insurance claims rules and to improve transparency of insurance and efficiency of claims.

Wujin Hospital and Zhenglu Town Health Center of Changzhou city launched Medical Information Framework to solve the problems of:

1. Information silos &

2. Data privacy security problems.

Jingdong Zhizhen Chain Medical Traceability Platform JDDigits Blockchain records each vaccine data and traces vaccine source to ensure:

1. Quality and safety of vaccine

2. Make the end user to inject safe vaccine.

3. Assure consumer's medical safety can be protected.

4. Ensures Patient privacy and safekeeping of medical records.

Benefits of Blockchain for Smart Transportation:

1. Help car-sharing data store on chain and solve the crisis of trust and collaboration among participants in the car sharing industry.

2. Introduce blockchain technology into the traditional carbon emission record system to promote energy saving and emission reduction.

3. Help optimize the intelligent transportation network and improve the efficiency of social operations.

Case 1: China Transportation Chain:

Launched by Ministry of Transport & 5 companies in city of Wuxi to solve problems of:

1. Traffic Congestion through better information sharing,

2. Construction of urban parking facilities,

3. Penalties for traffic violations, and

4. Information security of the Internet of Vehicles.

Case 2: Carbon Bank (Automobile) Public Chain Platform:

BYD (Carbon Bank integration platform), DNV GL (Certification). Veechain (Blockchain platform for automotive lifecycle management) are collaborating to integrate information of cars, large passenger cars, and other vehicles for on-chain storage.

Blockchain in e-governance – Smart city of Bhopal, India

Smart city of Bhopal, India is working actively with Somish Solutions, a pioneering start-up in leveraging Distributed Ledger Technology based Data Exchange Framework (DEF). DEF seamlessly integrates with existing systems & infrastructure and serves as a core component for multiple use cases of Governance on Blockchain.

DEF USE CASES IN GOVERNMENT A secure way for verifying citizen identity and authenticating common citizen certificates such as birth, death, marriage license etc. Citizen ID Verification & Certification

Land Registry Streamlining the land registry process while maintaining security, privacy and improving transparency of provenance and efficiency within the government.

Data Exchange: Approval-based data exchange and decision making between various government departments. Streamlines processes and leads to higher transparency in Government-to-citizen as well as Government-to-business interactions.

Legal Documents Tamper proof critical legal documents like wills and contracts with a complete history of changes made throughout the documents' lifetime.

Health Services Medical data linked to citizen IDs, stored securely on Blockchain, that provides need-based access to care providers.

SOMISH implemented a POC with Bhopal State to share information of emergency healthcare services on live basis over the Data Exchange Framework.

Blockchain can also help in a variety of Smart City applications to save lives and improve quality of life. The following case study showcases the use of Distributed Ledger Technology to seamlessly connect various actors in a smart city ecosystem to improve emergency healthcare response.

Collaborative Smart City Emergency Response for Smart Cities

Smart cities in India have implemented various 'smart' elements across several dimensions to automate various aspects of the lifecycle of a citizen's interactions with the Government. To facilitate smooth coordination between the citizens and the smart elements present in the city, ICCC (Integrated Command and Control Center) has been set up by all the smart cities.

However, the communication among all the smart elements is still lacking coordination especially, with respect to a timely and coordinated data sharing.

Faced with one of the highest numbers of traffic deaths across all the megacities of India at 157 per hundred thousand of population, Bhopal has sought to leverage Distributed Ledger Technologies to integrate various elements of its Emergency Response actors and their activities.

CHALLENGES IN EMERGENCY RESPONSE MANAGEMENT

Stakeholders Involved in Emergency Response Management

The current system lacks synergy within departments due to the following scenarios:

- Emergency services like the 108 ambulance service (operated under the National Health Mission) and departments like the police department exist and operate in silos
- There are no active communication links between the Ambulance service and the ITMS (Intelligent Traffic Management System) responsible for performing smart traffic management
- Without prior notice, hospitals are unable to prepare for emergency cases being brought to it by the ambulance

Fig 4.7 Challenges in Emergency Response Management

By introducing the concept of distributed data ownership, where each stakeholder owns only their data, DEF removes a Single Point of Failure and helps introduce trust among all stakeholders. Stakeholders are able to transparently share information with an immutable audit trail of each transfer The ICCC is able to streamline the city's Emergency Response service with other departments such as the traffic management system, hospitals, surveillance systems and the police.

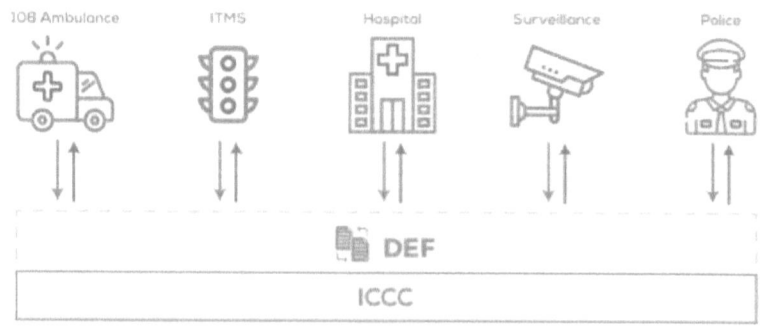

Fig 4.8 Data exchange framework of ICCC to connect all the players in the Emergency Response team (Source: Somish Solutions Ltd.)

By implementing a shared ledger by leveraging Distributed Ledger Technology, the Smart City of Bhopal's ICCC was able to create a rapid response to any emergency by the ambulance service to any accident case that is being reported by the citizens. The process flow is described below:

- The citizen reports an incident via the 108 Helpline.

- As the ambulance is dispatched the case details are shared with the relevant hospital, ensuring the staff has enough time to prepare for the emergency case.

- The ICCC is informed of the incident location and designated hospital, which activates the ITMS to create a traffic-free route leveraging the surveillance, public announcement and traffic management systems.

This establishes coordination among stakeholders, increased inter-department synergy and significantly faster transit time by triggering a green corridor for the ambulance transit.

The Smart City of Bhopal, in association with the Institute of Development Studies and National Institute of Urban Studies, successfully concluded the pilot in collaboration with Somish Solutions Ltd and is planning to expand the pilot for wider implementation.

Case study: Courtesy, Somish Solutions Ltd, New Delhi

Central Bank Digital Currency Projects

Another important Blockchain influenced application that can accelerate Smart city projects by facilitating digital value transactions between connected things and humans, while helping in unlocking the value of the data generated is the Central Bank Digital Currency Project.

Currently, the US Dollar is seen as the globally interoperable currency accepted by most nations. In the recent past, several countries are experimenting with the concept of leveraging the internet for speedy transfer of value considering the impending proliferation of IOT & Industrial IOT-led Home automation, Industrial automation and Smart City projects across the world. There has been a strong need felt for a digital equivalent of the national currencies giving rise to the concept of Central bank digital currency (CBDC), also called digital fiat currency (a currency established as money by Government regulation or law). Central Bank Digital Currency is different from virtual currency and cryptocurrency, which are not issued by the state and lack the legal tender status declared by the Government.

Various countries are already experimenting with the concept of CBDC and it is considered a transitory step to the ultimate eventuality of a fully digitized currency with the added security measure offered by a Blockchain approach.

According to the BIS, today some 70% of central banks are looking at CBDC, with most of them considering Blockchain as the underlying technology.

Some of the global Distributed Ledger Technology-based CBDC projects disclosed in the public domain are given in the following table published by Bank of Thailand in their project report on the state of CBDC project being experimented by BOT in conjunction with R3 Corda, Indian IT major Wipro and several transnational banks.

Phase	Paper Published	Project Focus	DLT Platform Used
Bank of Canada		**Project Jasper**	
Phase 1	Mar. 2016	1. Create a wholesale interbank RTGS proof-of-concept on DLT Ethereum platform 2. Evaluate PFMIs against tokenised interbank payments	Ethereum
Phase 2	Dec. 2016	1. Rebuild original proof-of-concept on Corda 2. Build additional functionalities such as LSM	Corda
Phase 3	Oct. 2018	1. Integrate a liquidity savings mechanism for netting transactions 2. Examine DvP solutions for security settlement	Corda
Monetary Authority of Singapore		**Project Ubin**	
Phase 1	Aug. 2016	1. Build a proof-of-concept for domestic RTGS on a private Ethereum network 2. Identify the non-technical implications of moving this into a production environment 3. Integrate DLT with existing RTGS in a test environment to automate tokenisation and detokenisation	Ethereum
Phase 2	Jul. 2017	1. Expand on the original proof-of-concept by incorporating LSMs 2. Understand how RTGS privacy can be ensured on DLT 3. Compare alternative DLT platforms	Quorum, Corda, Hyperledger Fabric
Phase 3	Nov. 2018	1. Explore different combinations of DLT for DvP between cash and Singapore government bonds 2. Test and examine solutions designed by Anquan Capital, Deloitte, and Nasdaq	Ethereum, Hyperledger Fabric, Chain, Quorum, Anquan
Central Bank of Brazil			
Phase 1	Aug. 2016	1. Identify use cases and build a working prototype for the central bank using DLT 2. Identify realistic functionality and build a minimum proof-of-concept for RTGS system on DLT platform	Ethereum
Phase 2	Nov. 2016	1. Analyse competing blockchain platforms using the selected use case as a benchmark 2. Address the privacy issues identified in the previous phase	Corda
European Central Bank & Bank of Japan		**Project Stella**	
Phase 1	May. 2017	1. Build RTGS system on DLT, including LSM functions 2. Assess safety and efficiency of current system in DLT implementation	Hyperledger Fabric
Phase 2	Nov. 2017	1. Build DvP proof-of-concept on different DLT platforms 2. Identify the trade-off between network size and performance 3. Assess DLT capability for cross-chain securities settlement	Quorum, Corda, Hyperledger Fabric, Elements
Hong Kong Monetary Authority		**Project Lionrock**	
Phase 1	Aug. 2016	1. Identify use cases and build a working prototype for the central bank using DLT 2. Identify realistic functionality and build a minimum proof-of-concept for RTGS system on DLT platform	Corda
South African Reserve Bank		**Project Khokha**	
Phase 1	Feb. 2018	1. Build an RTGS proof-of-concept on DLT exploring on privacy and scalability 2. Perform tests under a variety of deployment models in different locations 3. Assess a Quorum-based interbank payment system	Quorum

Fig 4.9 https://www.bot.or.th/English/FinancialMarkets/ProjectInthanon/Documents/Inthanon_Phase1_Report.pdf

Case Study: Central Bank Digital Currency experiment by Bank of Thailand (Project Inthanon) https://www.bot.or.th/English/ FinancialMarkets/ProjectInthanon/Documents/Inthanon_Phase2_ Report.pdf

Project Inthanon was devided into three progressive phases, each leveraging on the findings and learnings of the previous phase:

01 Phase 1 – Building the Fundamental

A POC for a DLT-based RTGS using wholesale CBDC for interbank settlement was built. A key highlight was the development of an innovative GR architecture with integrated Automated Liquidity Provision (ALP) functionality that achieved privacy and atomicity properties.

02 Phase 2 – Enhancing Functionalities

The objective is to build on the Phase 1 POC and augment it with additional functions to handle DvP settlement for interbank bond repo & trading, data reconciliation and handling of NR regulatory requirements. Outcomes from Phase 2 demonstrate the practicality of DLT at enabling transformative process improvement and technical feasibility of achieving DvP in real-time through an experimental MLSM.

03 Phase 3 – Exploring Cross-Border Funds Transfer Models

The DLT-based RTGS prototype will be expanded to connect with the other systems to support cross-border funds transfer transactions. The scope will also cover the regulatory and compliance issues from both THB and foreign currencies.

Fig 4.10 Project Inthanon

Bank of Thailand completed Phase-1 and Phase-2 successfully and has demonstrated many advantages of issuing the Central Bank-backed digital currencies over a Distributed ledger.

The findings from Phase-2 demonstrated the feasibility of Smart Contract utilization to automate bond life cycle events and DvP (Delivery Versus Payment) of inter-bank bond trading and repo transactions. The effective use of Smart Contracts has shown the potential to significantly streamline operational workflows and increase efficiencies. Fraud-prevention capabilities of the RTGS system were also augmented by the creation of a new end-to-end workflow that allowed validation of transactional information with external sources through integration points. The use of Smart Contracts for regulatory compliance purposes was also successfully tested in Phase-2 with the introduction of the NRFS mechanism, which could potentially eliminate multiple manual operational processes and allow banks to monitor NRBA/NRBS limits more effectively.

Land Titling using Blockchain in Smart Cities

A number of leading Digital Nations like UK, Dubai, Sweden have already started using Blockchain for maintaining Land Records, following case study of Land Pooling in APCRDA (Andhra Pradesh Capital Region Development Authority) elucidates how developing countries which are plagued by frauds and court cases in their Land Registry system can leverage Blockchain. Final implementation is however a long drawn out process involving multiple interactions, government transitions and resistance to change that could derail the process.

Case study of APCRDA for Blockchain in Land Pooling & Land Titling (Source: AP Govt records in public domain) is discussed here.

Back ground:

When a new city is developed from the scratch, land has to be pooled from existing owners and redistributed. In this process, the land parcels owned by individuals or group of owners are legally consolidated by transfer of ownership rights to the authority. It later transfers the ownership of a part of land back to the landowners for undertaking of development of such areas. In LPS the landowner will be getting Returnable Plots for the land he has surrendered with consent.

Implementation Process

Blockchain solution integrates with existing department systems through API calls & coexists non intrusively.

- Uses http API calls – can integrate with systems/GIS Systems on different technical platforms

- APCRDA GIS System data (Land Information) has been stored in Blockchain in Geo-json format.

- Modifications/Alterations of land records to follow the approved process and option is given for Authenticated Users only.

- Upon User request, GIS System (ArcGIS Server) generates Parcel images (Parcel, Block & Colony level location maps) along with Coordinates and Centroid of Parcel for Registration.

- Request API at Block chain server generates Block Chain Certificate embedded with QRCode (Information of Property).

- Existing systems include all transaction validation business logic & call APIs of Blockchain for respective data

Solution Architecture:

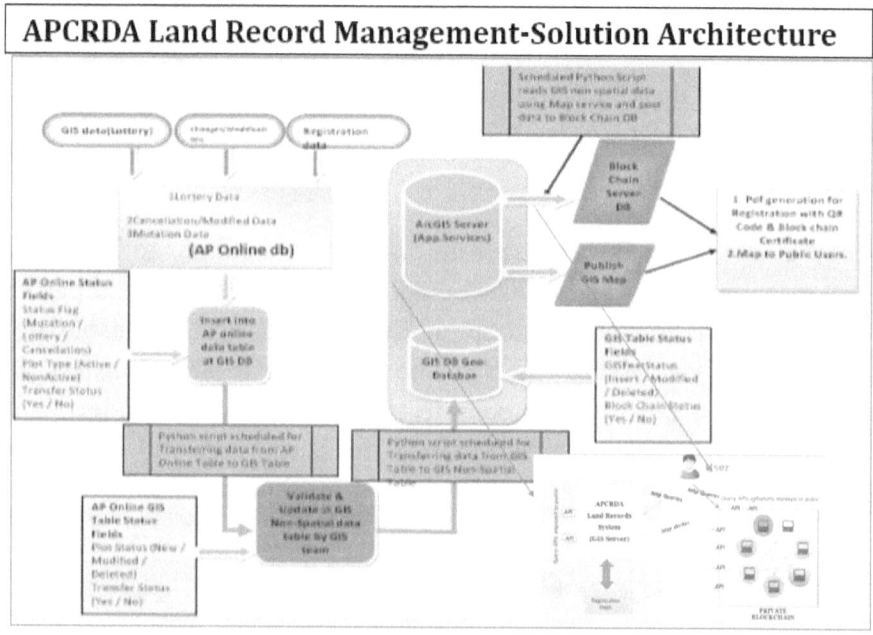

Fig 4.11 APCRDA Land Record Blockchain application Solution Architecture

Blockchain based Property certificate is issued to the owners and verifiable for authenticity on Blockchain.

Fig 4.12 Blockchain based QR code embedded Land ownership document

Blockchain and Sustainable Development Goals (SDG)

The Sustainable Development Goals are a collection of 17 global goals designed to be a "blueprint to achieve a better and more sustainable future for all." The SDGs, set in 2015 by the United Nations General Assembly and intended to be achieved by the year 2030, are part of UN Resolution 70/1, the 2030 Agenda. (Wikipedia)

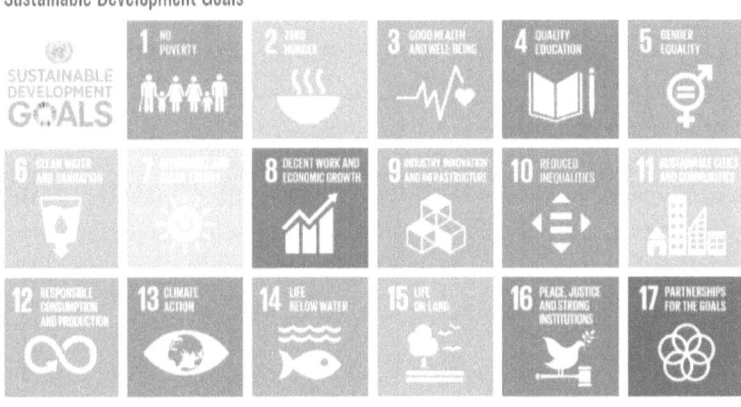

Fig 4.13 Sustainable Development Goals: Source: https:// sustainabledevelopment.un.org/sdgs

Blockchain has helped in a multipronged approach to attain the Sustainable Development Goals by all nations. The utility of Blockchain for the same is given as follows.

1. **Eliminate Poverty:** Targeted Govt. benefits to poor persons with no leakages.

2. **Eliminate Hunger:** Support Humanitarian activities targeted at food distribution in a coordinated manner. Crowdsource and track information on people deprived of daily minimum needs. World Food Program, an example.

3. **Good health and well-being:** Track medical records, eliminate fake drugs, support clinical trials and Pharma research and deliver and monitor high-quality subsidized cheap drugs. Track immunization health records of children through their early life. Insurance for all, especially pregnant women, elderly, poor and vulnerable tracked through Blockchain.

4. **Quality education:** Track academic credentials on Blockchain and support brilliant and downtrodden through scholarships and the right opportunities for global exposure of talent of them and to them.

5. **Gender Equality:** Incentivize and reward organizations and regions showing better performance on gender parity in areas like board rooms, staff ratio and Woman safety. Offer a channel for new employment opportunities for women with career breaks and with handicaps. Track safety measures and actions against atrocities for women in a coordinated manner.

6. **Clean water and sanitation:** Tracking effluents of industry, water pollution levels of major river bodies, the health of lakes, utilization of budgets targeted for Water conservation, Rainwater harvesting track records and efforts, etc.

7. **Clean renewable energy:** Enable peer-to-peer renewable energy trading, Reward renewable energy consumption, Track carbon certificates, facilitate measurement of usage and generation.

8. **Sustainable employment:** Verified expertise credentials and facilitate the gig economy for trusted peer-to-peer project marketplace

9. **Innovation, industrialization, and infrastructure:** Protect patents and help share and monetize intellectual capital. 3D manufacturing for productive industries and fast deployments, encourage recycling and reuse of industrial waste and residue.

10. **Reduce country-level inequalities:** Cross-country global cooperation, resource trading with reduced costs and complexities.

11. **Safe Smart Cities:** Secure IOT infrastructure with Blockchain for scalable automation.

12. **Responsible Production and consumption:** Tracking supply chains for ethical sourcing and providing live accurate data for forecasting.

13. **Climate action:** Track and reward environment conservation actions and progressive improvements across regions for reducing pollution.

14. **Life underwater:** Track the quality of seawater for harmful effluents and take steps to address deteriorations.

15. **Life on land:** Track forest fires on live basis across the world, take steps to track and improve afforestation, check desertification, aid in disaster management activities through coordinated actions.

16. **Peace, Justice and Strong Institutions:** ID2020 and Digital Identities. Blockchains enable trust which would, in turn, help mitigate corruption.

17. **Partnerships:** Global partnerships with win-win associations with collaboration, coordination, communication, cooperation facilitated by Blockchain.

Thus, Blockchain has significant applications in Smart city Projects and in achieving Sustainable Development Goals as laid down by United Nations.

Smart city projects are an important part of the Digital Transformation projects of Governments across the world.

In July 2014, OECD member countries, through the OECD Council, formally adopted a Recommendation **"that governments develop and implement digital government strategies"** to assist and guide them to achieve that digital transformation emphasising the role of digital technologies:

1. as a strategic driver to create open, participatory, and trustworthy public sector,

2. to improve social inclusiveness and government accountability, and

3. to bring together government and non-government actors for developing innovative approaches to contribute to national development and long-term sustainable growth.

The Recommendation also sets out a number of principles to guide the process of setting and implementing digital government strategies relating to engaging citizens and open government to maintain public trust; improving governance for better collaboration and results; and strengthening capabilities to achieve returns on investments in digital technologies. **(Source: OECD).**

D5 Nations namely Estonia, Israel, New Zealand, South Korea and UK have been the pioneers in adopting the Digital Transformation Strategies for the benefit of their people. A number of other countries have joined them over a period of time and a number of other leading countries like Singapore, USA, China, Dubai & UAE, Thailand, Switzerland, Germany, Australia and India have been following the OECD guidelines, to develop and work concertedly to leverage the emerging technologies to leapfrog the development through implementation of Smart City Projects.

In the subsequent chapters, we shall examine in detail how Singapore as a Digital Nation and the world's best SMART city has excelled in this aspect and stood out as a role model to the rest of the world.

Summary

This chapter threw light on the various risks associated with IoT devices and how Blockchain offers an excellent risk mitigation mechanism for IoT devices and Smart cities. The utility of Blockchain for Smart cities are manifold and some of them are:

1. Protecting Drones, Robots, Autonomous vehicles.

2. Manging land records, Health records, Property and educational records of citizens.

3. Help in Collaborative emergency response to save lives of accident victims.

4. Blockchain enables Digital currency can give fillip to M2M payments.

5. Catalyse the movement towards achieving Sustainable Development goals.

6. Emergence of D5 Nations to further Digital Transformation of Governance.

CHAPTER 5

Digital Transformation and Digital Government Strategies

Introduction

As per OECD, Digital transformation of Governments involves leveraging information and communication technologies for modernising public services, increasing service productivity and reducing labour intensity, increasing the level of satisfaction with and effectiveness of services, and increasing the openness of, trust in and engagement with governments.

In this chapter, we examine in detail the concept of Digital Transformation of Governments. We look at how the leading Digital Governments across the world are implementing strategic action plans to leverage advanced technologies for healthy, happy and comfortable life of their citizens thus, empowering them with all that the modern day lifestyles could offer.

The increasing penetration of information technology related products and services and their applications across all facets of our lifestyle, has led to the need for an integrated and coherent approach with a long term, synergistic strategy to weave together a platform, that facilitates the configuration, engagement and delivery of services. There is a need for accountability to ensure that all the services are in good shape, are handled securely, developed in alignment with the overall vision and continuously updated to keep up with the environmental development and customer tastes. If we see the way Apple Computer Inc launched iPoD in early 2000s and went on updating the product to different upgraded versions of iPhone with feature enhancements in line with the competition and consumer tastes, while serving as a platform to deliver myriad of services to their customers, one can appreciate the importance of Digital Transformation. Digital Transformation is a holistic concept that has enabled Apple Computer to keep in touch with the ecosystem to come out with new products and solutions all the time to survive and thrive in a technologically challenging marketplace.

Another important concept that is being leveraged by the digitally savvy successful organisations across the world is 'Digital Twin.'

Digital Twins

Digital twin refers to a digital replica of potential and actual physical assets (physical twin), processes, people, places, systems and devices that can be used for various purposes.

(Source: Wikipedia)

With the ubiquitous availability of Sensors and connected devices everywhere along with the enabling technologies like Cloud, Analytics, 3D modelling, Augmented/Virtual Reality and 4G connectivity etc., it is possible to not only possible to create a perfect simulation of any physical object or activity, but also track it on a live basis, even as we try to control its behaviour through a feedback loop.

In the context of Digital Built Britain, a digital twin is "a realistic digital representation of assets, processes or systems in the built or natural environment." Digital Twins of physical assets are helping organisations to make better informed decisions leading to improved outcomes. (Reference & Must read: https://www.cdbb.cam.ac.uk/system/files/documents/TheGeminiPrinciples.pdf)

The Gemini Principles

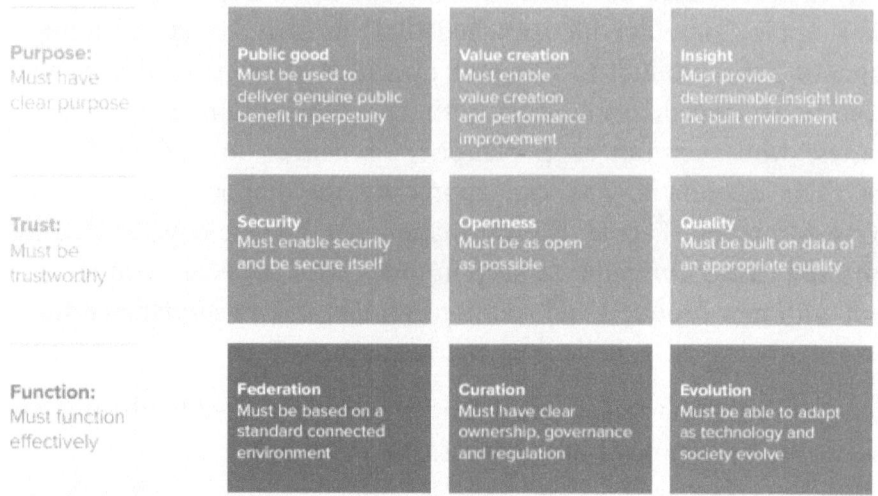

Purpose: Must have clear purpose	Public good Must be used to deliver genuine public benefit in perpetuity	Value creation Must enable value creation and performance improvement	Insight Must provide determinable insight into the built environment
Trust: Must be trustworthy	Security Must enable security and be secure itself	Openness Must be as open as possible	Quality Must be built on data of an appropriate quality
Function: Must function effectively	Federation Must be based on a standard connected environment	Curation Must have clear ownership, governance and regulation	Evolution Must be able to adapt as technology and society evolve

Fig 5.1a Gemini Principles – Value system behind the National Digital Twin platform – UK Govt

The Gemini Principles, a set of nine values to be embraced by all infrastructure leaders, are outlined by a study commissioned by CDDB (Centre for Digitally Build Britain) that is focused on creating a national digital twin (NDT) platform to unlock the value of digitisation across organisations & public utilities. This is expected to result in the use of "data for the public good" and result in better use, operation, maintenance, planning and delivery of national and local assets, systems and services.

Digital Twin concept has revolutionised every area of our operation and is a very important enabler in creating perfect architecture in the case of infrastructure and buildings.

For example, the Digital simulation of a building with all its surroundings and environmental conditions in the form of a 3D Simulated graphic, tracked on a real time basis and superimposed, enable Singapore government to create a perfect architecture for its millions of square feet of residential & commercial apartment and high rises to ensure high level of ventilation, free flow of breeze thus resulting in reduced energy consumption for air-conditioning and lighting. This is only one of the many possibilities, made available by the employment of cutting-edge technologies and a strategic mindset is needed to take advantage of this potential and make it work in a holistic manner. This is the hall mark of Digitally advanced nations and organisations.

Digital Governments adopt innovative and pro-active approaches to plan & execute Digital Transformation strategies in tune with the times.

Digital welfare is the application of Digital transformation strategies for digitisation of education, healthcare, and social care and protection services, including smarter use of well-proven assistive technologies with an objective of enhancing efficiency, effectiveness and good governance across the spectrum of activities in this domain.

	Information and Communication Technologies		
	Digitisation
Greater use of digital technologies to improve cross government activities and data/ information management	E-Government
Use by governments of digital technologies, particularly the internet, to achieve better government	Digital Government
Digital technologies and user preferences integrated in the design and receipt of services and broad public sector reform – integral part of governments' modernisation strategies to create public value
Change path	From a focus on: efficiency and productivity	Through a focus on: efficiency and productivity in delivering tailored services to individuals	To a focus on: governance, openness, transparency, engagement with and trust in government, as well as efficiency and productivity
	From Government-centred – users are passive recipients of services	Through User / Citizen-centred – users participate in service delivery processes	To People-driven – users voice their demands and needs, contribute to shaping the agenda and services' content and delivery
Public Services			
Administrative services
Internal core functions of government and internal activities in agencies that directly support service delivery	• Improve internal processes of government
• Improve internal processes supporting delivery of direct personal services, to improve services	• Innovative changes in internal processes
• Innovations in service delivery, at the margin	• Transforming internal processes
• Transforming service design and delivery
Direct personal services
Government services provided to address the personal wellbeing of citizens and support public policy outcomes	• Individual databases and information systems
• Standardised service delivery
• Standardised services	• Integration of IT systems and databases
• Collaboration
• 24/7 online services	• Data sharing: data/information crowdsourcing/data analytics
• Joined-up administrations – ICT platforms for sharing information, services and enhancing collaboration
• Innovative services tailored to individual needs / ubiquitous services (m-government)
			Digital Transformation

Source: OECD

Fig 5.1b Elements of Digital Transformation of Governments (Source OECD)

A digital government environment is as characterised in the highlighted part of the diagram – it is one that is largely user-driven, with users voicing their demands and needs and thereby contributing to shaping the government policy agenda and the nature of and means for receiving integrated direct personal services. Achieving Digital Government will, in some areas, require progression through a period of e-government, the middle stage in digital transformation

Digital Transformation (Definition and Theories)

Digital Transformation is the use of new, fast, and frequently changing digital technology to solve problems. It is about transforming processes that were non digital or manual to digital processes. One of the examples of digital transformation is cloud computing. (Wikipedia).

Some of the thought leaders and top notch global consultancies like MIT labs, Thought works and Gartner have studied the best practices of the organisations that are undisputed exponents of bleeding edge technologies and are leaders in their respective domains to articulate the characteristics of well developed & effective Digital Transformation strategy,

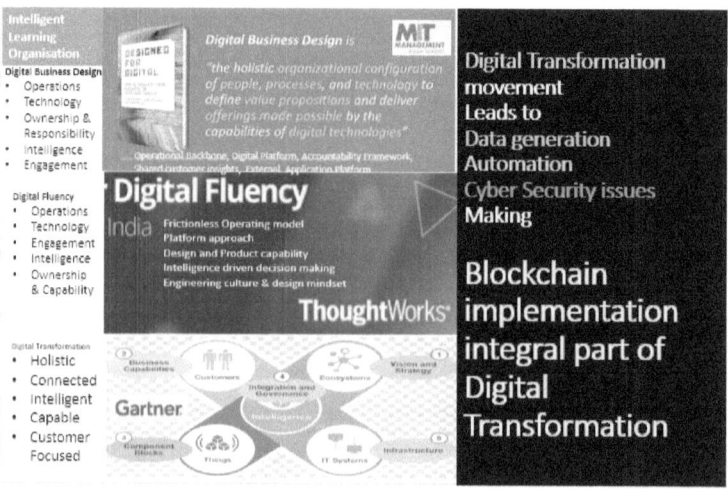

Fig 5.2a Elements of Digital Transformation and the critical place in it for Cyber Security & Blockchain

There have been many countries that have now embarked on Digital Government strategies. Here, we take a look at some of the leading countries that are a part of the founding group of D5 Nations (Now D10).

A study that has been conducted for analysing the success factors of Digitally advanced organisations, that have conquered the markets they are operating in, has outlined the following characteristics:

1. Operational excellence achieved through a sturdy framework & robust internal systems aligned well.

2. Digital platforms to connect the employees and engage the customers as well as partners that enables to seamlessly configure new products and services, while serving the stakeholders.

3. Customer centricity and use of advanced technology solutions to keep a pulse on their markets and speedily respond.

4. Strong accountability framework where every objective and goal chased is handled in a responsible manner to reach their logical conclusion.

5. An integrated and strategic approach that synergises all elements of physical, digital and technology aspects & aligns the activities of all stakeholders to achieve the vision laid out. In other words, every member of the organisation has been sold on the 'Purpose' of the organisational existence that leads to delivery of the products and services (what) in the best possible quality (how).

Digital Transformation Excellence Model

The Digital Transformation excellence model summarises the various elements and pillars (ABCD) of Digital Transformation in the following succinct way.

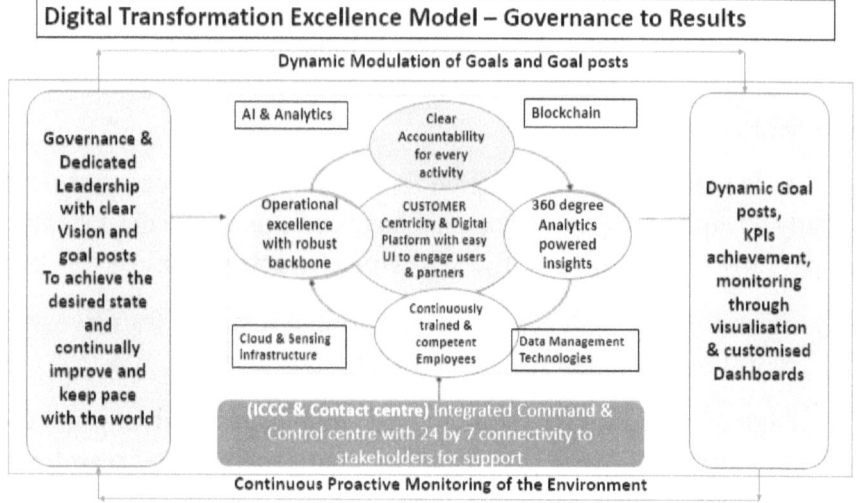

Adoption of Malcolm Baldridge Business Excellence Model to achieve Digital Transformation
(Illustration by Srinivas Mahankali)

Fig 5.2b Digital Transformation from Vision to Goals to Achievement to Renewed Mission

The various elements of the Digital transformation as discussed through the various models can best be illustrated using the Malcolm Baldridge Business models as shown above.

ABCD, the four key pillars of Digital Transformation are AI & ML powered analytics, Blockchain, Cloud enabled scalability & Sensor enabled collection of data & Data driven approach encompassing all data storage, warehousing, mining and visualisation technologies. All these combines brilliantly to derive the value from the Digital Transformation to achieve the desired goals.

Digital Nations

Digital 9 Network and its Objectives

The Digital 9, or D9, is a network of the world's most advanced digital nations with a shared goal of harnessing digital technology and new ways of working to improve citizens' lives

Digital Nations - Harnessing Digital Tech for improving Citizens Lives			
Country	Joining Month	Year	Status
Estonia	December	2014	Founder
Israel	December	2014	Founder
New Zealand	December	2014	Founder
South Korea	December	2014	Founder
United Kingdom	December	2014	Founder
Canada	February	2018	Member
Uruguay	February	2018	Member
Mexico	November	2018	Member
Portugal	November	2019	Member
Denmark	November	2019	Member

Fig 5.3 Digital Nations Group D9

Digital Nations signed a common charter at Uruguay in 2019 to enable the Digital Nations to support each other consists of the following areas:

1. User needs – the design of public services for the citizen;

2. Open standards – a commitment to credible royalty-free open standards to promote interoperability;

3. Open source – future government systems, tradecraft, standards and manuals are created as open source and are shareable between members;

4. Open markets – in government procurement, create true competition for companies regardless of size. Encourage and support a start-up culture and promote growth through open markets;

5. Open government (transparency) – be a member of the Open Government Partnership and use open licenses to produce and consume open data

6. Connectivity – enable an online population through comprehensive and high-quality digital infrastructure;

7. Digital skills and confidence – support children, young people and adults in developing digital competencies and skills;

8. Assisted digital – a commitment to support all its citizens to access digital services;

9. Commitment to share and learn – all members commit to work together to help solve each other's issues wherever they can.

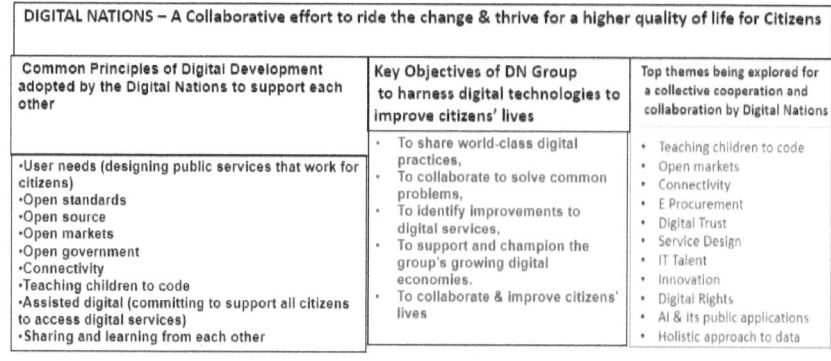

DIGITAL NATIONS – A Collaborative effort to ride the change & thrive for a higher quality of life for Citizens		
Common Principles of Digital Development adopted by the Digital Nations to support each other	Key Objectives of DN Group to harness digital technologies to improve citizens' lives	Top themes being explored for a collective cooperation and collaboration by Digital Nations
•User needs (designing public services that work for citizens) •Open standards •Open source •Open markets •Open government •Connectivity •Teaching children to code •Assisted digital (committing to support all citizens to access digital services) •Sharing and learning from each other	• To share world-class digital practices, • To collaborate to solve common problems, • To identify improvements to digital services, • To support and champion the group's growing digital economies. • To collaborate & improve citizens' lives	• Teaching children to code • Open markets • Connectivity • E Procurement • Digital Trust • Service Design • IT Talent • Innovation • Digital Rights • AI & its public applications • Holistic approach to data

Fig 5.4 Goals and Objectives of Digital Nations

The Finest Digital Nations

Let us look at some of the founding nations of D9 & their Digital Transformation strategies.

Estonia

A small country with 1.3 million population, Estonia is one of the most advanced digital societies. It has built a strong & integrated digital infrastructure that offers a safe and seamless e-services ecosystem. The system offers a flexibility and the ability to integrate its different parts, while improving e-services and allowing government systems to

Fig 5.5 Estonia's e-Ecosystem

grow. Seamless digital services increase the potential for economic growth and a higher quality of life.

Estonia uses one of the most advanced Digital Citizen Identity management system.

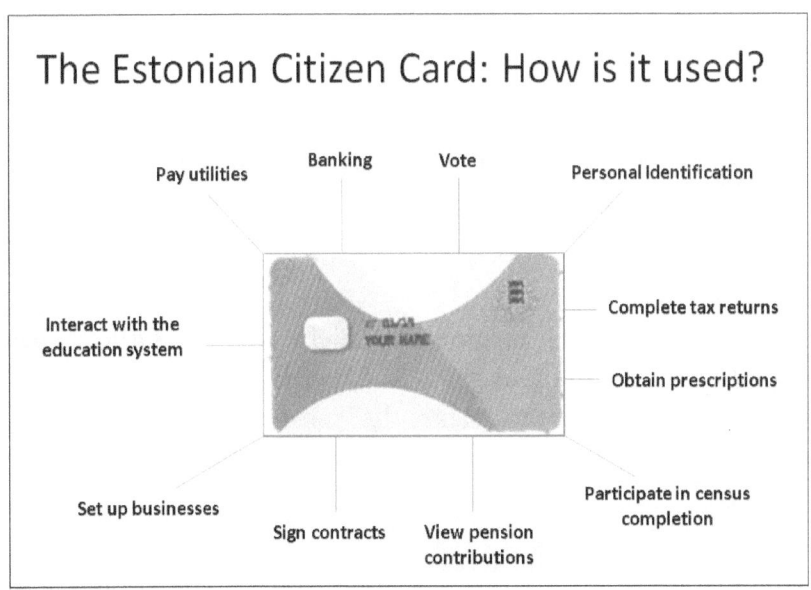

Fig 5.6 Estonia's Citizen identity card (Source: Estonia Government portal)

Through their identity card, online access is provided for 100% of citizen Government services and for businesses, multiple services such as registering businesses and properties and applying for social benefits. Certain benefits such as family benefits are even triggered automatically by events such as the birth of a child and its registration. Secure exchange of data and authentication is done electronically in a seamless manner. Official decisions are confirmed with a digital stamp, and individuals can sign with digital signatures. These digital versions are equal to physical stamps or signatures under Estonian law.

Digital Health records and e-services as mentioned earlier enable Estonia's citizens to avail contactless citizens during the trying times of the Pandemics. (Source: https://www.weforum.org/agenda/2020/07/estonia-advanced-digital-society-here-s-how-that-helped-it-during-covid-19)

New Zealand

(https://www.digital.govt.nz/digital-government/digital-transformation/nz-digital-transformation/)

Focus of New Zealand's Digital government strategy is to leverage emerging technologies to meet the citizens' needs using emerging technologies, data and fine tune Government's practices and processes.

New Zealand put in place a team of 55 senior leaders from more than 20 agencies working together through the Digital Government Partnership to support the goal of a coherent, all-of-government digital system.

New Zealand's Government Chief Digital Officer works collaboratively with the Government Chief Data Steward and leaders across the state sector to make the most of digital technologies to transform the way government works for the benefit of New Zealanders.

The objectives of the Program are to ensure everyone shares the benefits of digital transformation, which will see many aspects of government and ultimately the lives of New Zealanders significantly improve that as follows:

- work with people and other organisations to design services;

- take advantage of the opportunities new digital technologies present and prepare people for the changes these technologies bring to our society and economy;

- put the right systems, settings and infrastructure in place;

- ensure our workforce has the right skills to succeed in a new environment;

- enable innovation and experimentation to happen freely;

- provide better access to data, while making sure the privacy of individuals is protected;

- support people who can't access the digital world and

- invest smartly.

Activities undertaken by The Digital Government Team of New Zealand:

- Developed and launched multi-agency services based around key events in people's lives, like having a baby;

- Created the world's 1st fully online passport renewal service—and we've extended it to new passports too;

- Developed the Data Investment Framework for a more efficient and consolidated approach to investments in data infrastructure;

- Established the Government Chief Privacy Officer to help government agencies manage privacy and security more effectively. Privacy, security and risk;

- Changed the way the New Zealand Government invests in ICT by creating shared ICT capabilities—this has resulted in savings and cost avoidance of more than $100 million each year;

- Developed a system-wide view of ICT assurance and risk. Privacy, security and risk and

- Made it easier for people to find information in 1 place on Govt.nz, regardless of the agency the information belongs to Govt.nz.

The government provides detailed advice and guidance to help people plan and develop digital services and systems. This helps us maintain consistency across agencies, share our knowledge and promote best practices so the end result is a better experience for everyone who interacts with government.

Republic of South Korea

South Korea has been one of the most advanced countries in terms of digital governance for many years now. Its telecommunication infrastructures development rate, online services, and citizens participation online are high.

The Korean government announced a basic plan for the intelligent government in March 2017. This was because the administrative environment for implementing the digital government had been largely affected by intelligent information technology. Specifically, the combination of digital data and artificial intelligence technology made it possible to improve the rationality of the administration, scientific quality, and customized services according to area, class, and situation. Korea had great potential because of the accumulation and use of large amounts of digital data in administrative areas through continuous e-government promotion.

On July 19, 2017, the National Planning and Advisory Committee selected 20 national strategies and 100 national agendas and announced a five-year plan for national vision and state administration. The presidential decree on the creation and management of the Presidential Committee on the Fourth Industrial Revolution (PCFIR) was promulgated and came into effect on August 22, 2017. The PCFIR was established in response to the rapid changes in technology in the face of the fourth industrial revolution to address issues and policies related to new and advanced science and technology (e.g., artificial intelligence, big data).

The highlight of the Korean experience is the understanding that from the standpoint of digital government transformation using information technology, it is most important to promote digital government policy directly from the ministry that manages the budget, or to establish a dedicated organization under the ministry to secure strong coordination while linking it with the budget.

Reference: A Comparative Study of Digital Government Policies, Focusing on E-Government Acts in Korea and the United States – DOI: 10.3390/electronics8111362: Jour Chung Kim 17/11/2019

United Kingdom

The UK Government is one of the most digitally advanced in the world. We have come top of the 2016 United Nations E-Government and E-Participation surveys. UK has developed the award-winning and internationally renowned GOV.UK – and opened its code which has been reused by governments around the world.

Digital Transformation in United Kingdom government is driven by, Government Digital Service (GDS), a part of the Cabinet Office. GDS, aims to help all of government use emerging technologies in the right way to address the right problems

The UK digital government innovation efforts, under the leadership of the GDS, focused on reestablishing government–citizen relations so that the digital government can react more swiftly to the surrounding environment to provide citizens with greater authority. GDS, which is leading the UK digital government innovation, can also strongly promote its policy as a Cabinet Office organization

It functions as a Centre of Excellence in digital, technology and data, collaborating with departments to help them with their own transformation and to build platforms, standards, and digital services.

GDS has been viewed as a model by many developed countries like USA, Australia and New Zealand.

Some of the bottlenecks that must be addressed as identified in interaction with Private sector innovators are:

- a need to build capability and an innovation culture among civil servants

- procurement making it hard for innovative startups to work with government

- issues around combining legacy infrastructure with new technology

To address these, GDS has undertaken a number of initiatives, like building the Digital Market Place. Some of the initiatives and activities are outlined below:

GDS built and runs GOV.UK, the best place to find government services and information a key part of our national digital infrastructure.

GDS works on a number off Digital transformation activities with a mandate to:

- work with the rest of government to make public services simpler and better

- build platforms like GOV.UK Verify – a way to confirm users are who they say they are

- work to ensure government data is good data, more usable for all

- help departments make better-informed decisions when they need to buy technology

- help departments provide their staff with better value technology that's more of a tool and less of a barrier

The next stage of digitally-enabled transformation has three broad components, which together form the scope of this strategy:

i. transforming whole citizen-facing services to continue to improve the experience for citizens, businesses and users within the public sector;

ii. full department transformation-affecting complete organisations to deliver policy objectives in a flexible way, improve citizen service across channels and improve efficiency; and

iii. internal government transformation, which might not directly change policy outcomes or citizen-facing services but which is vital if government is to collaborate better and deliver digitally-enabled change more effectively.

Reference: https://assets.publishing.service.gov.uk/government/uploads/system/uploads/attachment_data/file/590199/Government_Transformation_Strategy.pdf

Israel

The national initiative "Digital Israel," whose bureau is in the Ministry for Social Equality, is a government project aimed at harnessing the potential of the digital revolution and the advancement of information and communications technologies (ICT) to accelerate economic growth, reduce socio-economic gaps and transform the government into a smart, efficient and citizen-friendly world leader in the digital domain.

The vision of the program is to leverage the potential of the digital revolution to help reduce social and geographic gaps, accelerate economic growth, and promote a smart and friendly government that will position Israel as a world leader in digital governance. The National Digital Program was written considering this vision and reflects the Israeli government's digital policy for the years 2017-2020. In order to realize the vision of the national initiative, the overall goals and strategic objectives of the program have been defined.

The activity scope of the National Digital Program, are depicted in the following figure:

Fig 5.7 Activity scope of National Digital Program, Digital Israeli (https://bit.ly/31AzwMQ (Israel National Digital Program)

The National Digital Program was designed to promote the initiative's goals and scope of activity, and to serve as a digital compass for all ministries and government bodies. The program is a framework for the government's activity in implementing the national initiative and the ministries and support units will work to realize the initiative, either through ministry work plans or through special digital ministry programs, assisted by the Digital Israel Bureau. In order to assist government ministries in formulating a digital strategy, the program includes several guiding principles: targeting customer needs, information resource management, digital inclusion and digital by default. These principles are consistent with the government's ICT strategy formulated by the Government ICT Authority. The program will help build growth engines for the economy and create national benefits for the state and for society, and its realization will give Israel the opportunity to make a significant and necessary digital breakthrough. (https://www.gov.il/en/departments/digital_israel)

Case Study – Integrating ICT across welfare policy areas in Denmark: The Common Public Strategy for Digital Welfare 2013-2020: Empowerment, Flexibility and Efficiency

"The Common Public Strategy for Digital Welfare 2013-2020" is an important pillar in the Danish Government's medium-term planning framework "Growth Plan Denmark." In this plan, modernisation of the public sector is expected to free up resources corresponding to 12,000 million DKR (1,600 million Euros) in 2020.

This aim means that while a number of themes are handling public service transformation, increasing and developing qualities, the perspective of and focus on realising efficiency gains remains an underlying precondition for the projects of the strategy.

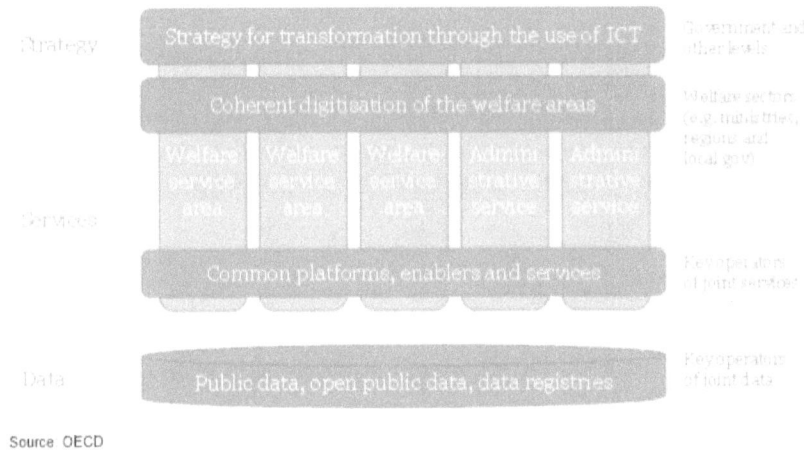

Source OECD

Fig 5.8 A whole of Government approach to Digitisation

The Strategy is joined-up across all levels of government, complementing the existing e-government strategy with a focus on further digitisation of public welfare services. The strategy addresses seven focus areas:

- National roll-out of tele-medication, including identifying relevant areas, testing new patient groups, and providing the necessary infrastructure.

- Effective collaboration in the healthcare area, including digital booking at hospitals, better user of patients' own information, implementation of a joint national medication card, fully digital communication in the healthcare sector, increased use of video conferencing, for example for interpretation.

- Welfare technology and care, including the roll-out of devices to help lift patients, washing toilets, use of eating robots in senior housing facilities, digitally supported recovery and testing of smart homes.

- New digital approaches in case handling, including freeing up resources through speech recognition, better evidence in social policies and programmes, and increasing quality through better data sharing;

- Digital learning and education, including using digital teaching aids and educational materials in schools, digital exams, digital tools for day care;

- Digital cooperation in education, including a joint user portal for primary schools, a digital folder to store all educational certificates, better sharing of digital learning tools;

- Preconditions for digital welfare, including sufficient broad band coverage, establishment of a joint public solution for mobile security, joint security standards and digital competencies.

The strategy outlines the key challenge of achieving a joint and coherent direction across the different public welfare policy areas, while at the same time establishing and maintaining a committed, de-centralised implementation capacity. Source: The Common Public Strategy for Digital Welfare 2013-2020: Empowerment, Flexibility and Efficiency, see www.digst.dk.

Summary

This chapter took a detailed look at Digital Transformation of Governance and its implementation in leading countries across the world. The following concepts were examined:

1. Digital Transformation definition and its elements as per various theories

2. Digital Twins and how this concept is leading to gainful digitisation

3. Digital Transformation Business Excellence model followed by leading organisations

4. Digital Nations, their themes and goals

5. Best Digital Nations in the world and what differentiates them

CHAPTER 6

Case Study of Singapore

CHAPTER 6.1: Introduction

Singapore rose from being a poor third world country to a flourishing & rich economy over the past 5 decades.

Divorced from a merger with Malaysia, in 1965, Singapore did not have anything to brag about!

The young impoverished nation with no own trained military while being at risk of military conflicts from all around was battered by problems like limited Natural resources including fresh water, illiterate population, over 15% unemployment rate, no proper housing for most citizens

Today, Singapore is a country with close to 6 million persons and the 3rd highest in the world in population density can be proud of,

1. Healthiest population

2. Near 100% literacy,

3. 6th in world in Per capita income,

4. #1 in ease of doing business,

5. Top 3 in the world in exploring & exploiting Emerging technologies and

6. Being the most sought-after International destination with excellent state of the art Seaports & Airports!

Singapore is one of the few countries in the world, where you do not find any sort of Religious, Racial, Economic discrimination with people living in utmost harmony, enjoying maximum freedom, while feeling secure! Singapore is a neutral country and maintains friendly relations with all countries across the world and its Passport is the most coveted, offering visa-free entry into over 150 countries.

Ranked on the top for being a business and innovation country, Singapore city is developing Punggol Digital District and the Jurong Innovation District as 'Innovation Districts' for fostering new age industries leveraging emerging technologies and Industry 4.0. Its one-north district that houses Block 71, is home to more than 500 start-ups.

Hyundai Motor Company is establishing Hyundai Mobility Global Innovation Center in Singapore (HMGICs) to accelerate its innovation efforts, with support from the Singapore Economic Development Board (EDB) in Singapore's Jurong Innovation District, a one-stop advanced manufacturing hub developed by JTC Corporation, housing a vibrant ecosystem of researchers, technology and training providers and factories of the future. This is expected to support Hyundai's transformation into a smart mobility solution provider, while positioning Singapore firmly as a preferred destination for innovation and entrepreneurship globally.

Visionary Leadership Across the Decades

The Right Honourable
Lee Kuan Yew
GCMG CH SPMJ
李光耀

Lee Kuan Yew in 2002
1st Prime Minister of Singapore
In office
5 June 1959[1] – 28 November 1990

For this, Singapore under the able leadership of its early founders led by its visionary Prime Minister (Late) Lee Kuan Yew fought a determined battle to not only survive, but also thrive.

Lee Kuan Yew, one of the most visionary leaders of our times and the first Prime Minister of Singapore can be credited with the amazing and the most successful social engineering feats in human history.

Lee turned a chaotic and malaria-ridden harbor in Southeast Asia into a shining city state with one of the highest per capita GDP in the world.

Singapore's National Flag

Consensus and Thoughtfulness with a futuristic approach runs in every step and action this young nation takes. The same thing goes in the way the National Flag of Singapore evolved.

Adopted in 1959 and reconfirmed in 1965, on August 9, Singapore's National Flag consists of:

1. Horizontal bicolor of red and white; charged in white in the canton with a crescent facing the fly and

2. Pentagon of five equal sized stars resembling the nation's ideals with equal importance.

The elements of the flag denote a young nation on the ascendant, universal brotherhood and equality, and national ideals, Democracy, Peace, Progress, Justice and Equality.

Building Singapore as an attractive destination for leading companies from across the world, maintain friendly relationships with all the countries, following prudent policies of investment & attracting the best of talent from across the world to feel at home in Singapore and contribute, Singapore unleashed a flurry of productive and purposeful activity channelized in the right direction.

In 55 years, a barren land of impoverished people has been transformed into a heaven on earth with appropriate combination of residential, commercial, recreational, and magnetic tourist places. Most of the citizens from being homeless, are now proud owners of their own house, even as they are unable to afford purchasing their own four wheeled automobiles, thanks to the Government move of taxing the automobiles heavily. This has also resulted in well maintained traffic load on the infrastructure.

The activities of Singapore across various dimensions of sustainable infrastructure planning, technology for improved quality of life across all dimensions that matter for a citizen's life and for reaching self-sustenance in water and food have been commendable. We shall look at some of these dimensions in the subsequent chapters.

CHAPTER 6.2: The Leadership – Calm, Collected, Empathetic

Singapore's Leadership principals are ingrained into the thinking of its leaders, strengthening in their value systems as they are handed over from generations of experience and striving for excellence.

The hall mark of Singapore Leadership that can be observed from every bit of their practice & communication can be summarised as follows:

1. Be Self-confident, Practice Humility & Respect every human, valuing their life to the hilt;

2. Have a strategic & holistic perspective honed by a long-standing vision. Sustainability for Perpetuity is the key;

3. Encourage Experimentation and Use failure as a stepping stone for future success;

4. Encourage innovation & focus on Practical use cases that result in tangible benefits;

5. Focus on building strong foundations – Trust, Technology, Talent, Teamwork, Timely action, Tourist attractiveness, Travel infrastructure. Be the best and become a magnet for the best in the world

6. Focus on building relationships, developing networks and collaborate for success while being independent and

7. Strive continuously for sustainability, self-reliance & excellence.

The motto of the leadership generation after generation can be summed up in the following points:

- Be Self-confident,

- Practice Humility & Respect,

- Value every human life to the hilt and

- Focus on doing what is right while keeping the evil at bay.

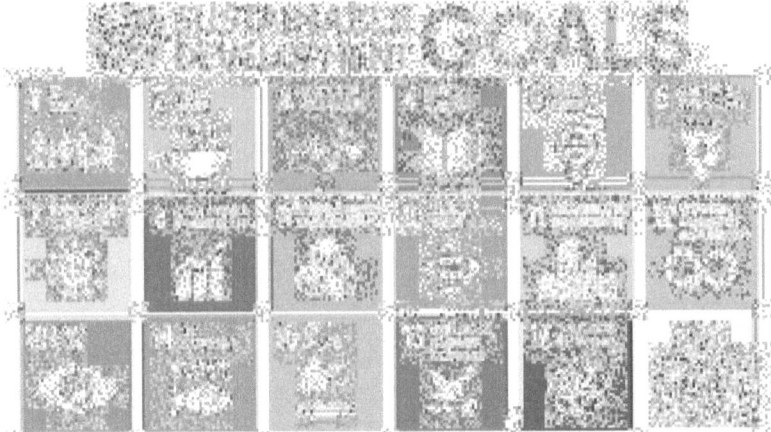

Singapore's leadership at the time of writing this book is shown in the following pictorial.

Fig 6.2.1 Singapore leadership team as of June 2020 that battled the Covid virus

I started closely following the actions and activities of the Singapore leaders for 5 months from January 2020 end to June end as they battled the most horrible crisis of our time, the nCovid19 virus.

Everything that Singapore government did to protect its citizens, caution them, serve them, treat them to save their lives & livelihoods seemed calibrated, measured & executed to perfection. Even when the things seemed to be going out of control from the outside, they seemed confident and in control of the situation. Leveraging technology, financial resources, distributing masks & kits, and above all an attitude to serve the people to the best, ensured that no lives are unnecessarily lost and the citizens' & the migrant workers' confidence in their Government seemed perfectly in place. The death rate of under 0.1% is astounding even to the best of the governments across the world.

PM Lee Hsien Loong comes across as a person of utmost humility, treating his Honourable President & his teammates, with utmost respect irrespective of the gender, age, background & position, the Singaporean way! Further, the Government seemed to be having the 'Right people at all the Right Places!'

CHAPTER 6.3: The 5T Magnet where Innovation Flourishes

The various pillars of Singapore's economy that attract everyone's attention like a magnet, can be well summarised in 5 key words namely Trust, Technology, Talent. Tourism and Trade

Trust – Through a National digital identity, citizen consent architecture offering complete data privacy & security, the Govt. gains utmost trust of its citizens.

Technology – Strategic focus on Open API driven approach, Blockchain, Cloud, AI/ML & Data analytics, Cybersecurity makes it world's leading destination for emerging technologies.

Talent – Encouraging entrepreneurship, focus on technology literacy from childhood to ripe old age, regulatory sandbox that fuels experimentation with investments, Singapore has turned into a magnet for global top-notch talent.

Tourism: Singapore sports many attractive tourist & recreational spots for all age groups & continues to be one of the topmost favourite destinations for travellers across the world.

Trade: Focus on ease of doing business, top notch sea, road, and airport infrastructure & friendly government support builds on its locational advantage. Leveraging Blockchain for secure digitisation will leapfrog Singapore's trade attractiveness.

Collaboration and Cooperation is a must for survival. Protectionism and parochial policies leading to silos owing to the Protectionist regimes are a bane to global development. To ward off this threat, Singapore participated in mutually beneficial trade agreements with several like-minded and progressive countries. Singapore signed Comprehensive and Progressive Agreement for Trans-Pacific Partnership ("CPTPP"), a Free Trade Agreement ("FTA") between 11 countries in Asia-Pacific to strengthen the trade resulting in a seamless flow of goods, services, and investment. Further it is also working on (Regional Comprehensive Economic Partnership) agreement with 15 countries contributing to over 35% of world's GDP. Thus, Singapore is building on its strengths and collaborating with the rest of the world to create productive trade networks that strengthen its position as a global trade hub.

Ranked #1 in the world for ease of doing business, Singapore has many incentives for budding entrepreneurs that allows the innovation to be commercialised in a speedy manner.

All the entrepreneurs are well encouraged to propose their new ideas to a seasoned set of Public-Private mentor networks and the selected ideas are put through a well-designed Sand Box system with immense Government support, to examine viability & potential of the envisaged solutions. Once the ideas are proven in the Sandbox, they are immediately primed for implementation with the full force of Government behind them as a facilitator, adopter, funder, and implementer.

How does the Regulatory Sandbox Work?

Case study: Regulatory Sandbox – EMA (Energy Metering Authority, Govt of Singapore)

(https://www.ema.gov.sg/media_release.aspx?news_sid=2017 1020Wab84AqS9NXY)

1. Regulatory Sandbox enables the energy sector to test new products & services in the electricity & gas sectors, before deciding on the appropriate regulatory treatment.

2. This is designed to help the innovators to leverage on new technology or apply existing technology in novel ways to create value for electricity and gas consumers, or to improve business and operational procedures, without the risk of a major failure that normally stifle such innovations

3. EMA is encouraging innovators to apply for such experimentation by enabling such ideas to be tested through a Regulatory Sandbox. A successful application would allow the idea to be applied in the market, while being subject to relaxed regulatory requirements, in a controlled environment that limits the risks to consumers and industry.

4. The evaluation criteria for ideas applying for the Regulatory Sandbox include whether the proposal:

    i. Uses technologies/products in an innovative way;

    ii. Addresses a problem or brings benefits to consumers and/or the energy sector;

    iii. Requires some changes to existing rules; and

    iv. Has assessed and mitigated foreseeable risks.

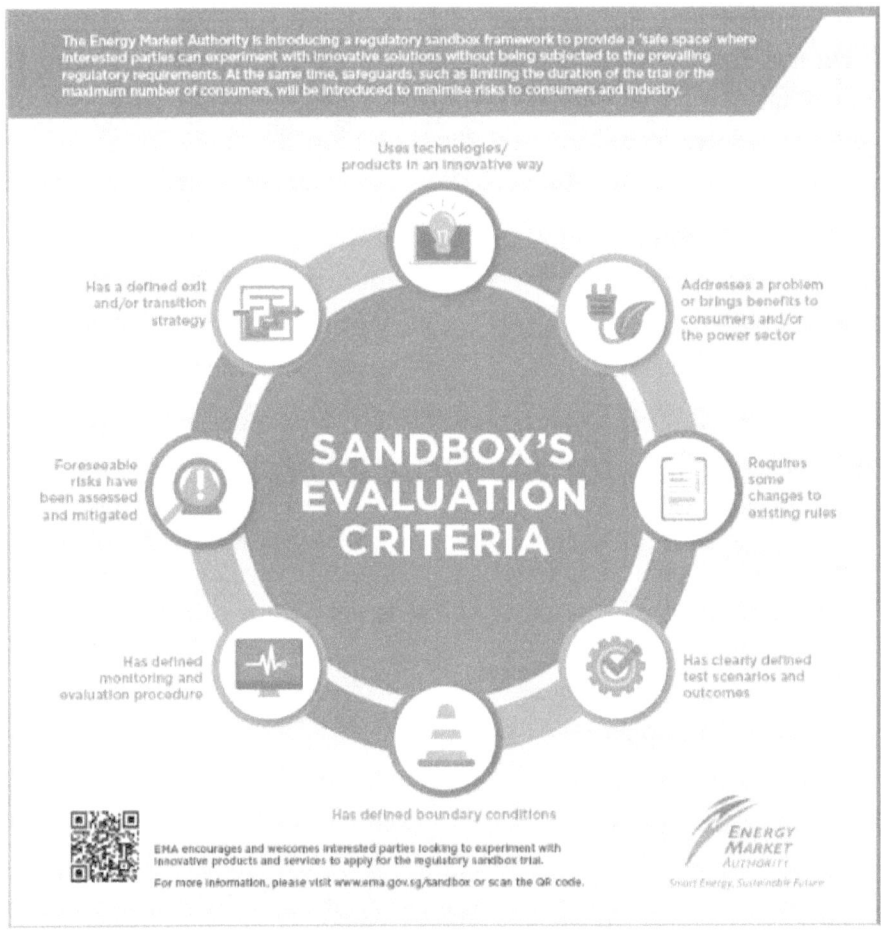

The Energy Market Authority is introducing a regulatory sandbox framework to provide a 'safe space' where interested parties can experiment with innovative solutions without being subjected to the prevailing regulatory requirements. At the same time, safeguards, such as limiting the duration of the trial or the maximum number of consumers, will be introduced to minimise risks to consumers and industry.

Uses technologies/ products in an innovative way

Has a defined exit and/or transition strategy

Addresses a problem or brings benefits to consumers and/or the power sector

SANDBOX'S EVALUATION CRITERIA

Foreseeable risks have been assessed and mitigated

Requires some changes to existing rules

Has defined monitoring and evaluation procedure

Has clearly defined test scenarios and outcomes

Has defined boundary conditions

EMA encourages and welcomes interested parties looking to experiment with innovative products and services to apply for the regulatory sandbox trial. For more information, please visit www.ema.gov.sg/sandbox or scan the QR code.

ENERGY MARKET AUTHORITY
Smart Energy, Sustainable Future

Fig 6.3.1 EMA Regulatory Sandbox

5. The Regulatory Sandbox will help EMA adjust its regulatory frameworks to keep pace with advances in technology and enable promising innovations to prosper.

6. The Regulatory Sandbox will complement ongoing Energy Research and Development (R&D) initiatives, by providing a platform for R&D projects to be tested on a broader scale in Singapore.

This system attracts the best organisations, entrepreneurs and investors in the World that further fuels the innovation with high quality and chance of success. With focus on the ABCD of today's disruptive technologies namely API economy, Blockchain, Cloud & Data Analytics enveloped in a highly guarded Cyber security focused approach, Singapore leads in the adoption of these technologies towards a better quality of life for its citizens and the global population!

Today Singapore continues to thrive, guided by the values of openness, inclusiveness, self-determination, meritocracy, and incorruptibility.

Technology at Work-State of the Art Public Infrastructure for Singapore Citizens

Singapore is having a state of the art public digital infrastructure with comprehensive digital stack for financial services to operate and digitize enabling the FIs to perform the following operations

A/C Opening Via (MyInfo – ekyc)

A/C Access Via (SingPass)

A/C Transfer Via (PayNow)

A/C consolidated across FIs Via (API Market Place)

The ecosystem has a healthy combination of FinTechs, FIs and Public Digital Rails engaged optimally.

SingPass: Launched in March 2003, Singapore Personal Access (or SingPass) allows users to access hundreds of government services easily and securely online using their fingerprint or a 6-digit passcode.

SingPass Mobile is a mode of two-factor authentication (2FA). It involves:

1. The app that is in mobile phone (what citizen has) and

2. Your fingerprint (what you are), or 6-digit passcode (what citizen knows)

To protect the user's account, he/she has to register their own fingerprint on their smartphone and maintain secrecy of their 6-digit passcode. There are also several layers of security measures in place. For example, when the app detects potential security breach or presence of malicious software on your mobile device, it will not allow them to use the app on that device.

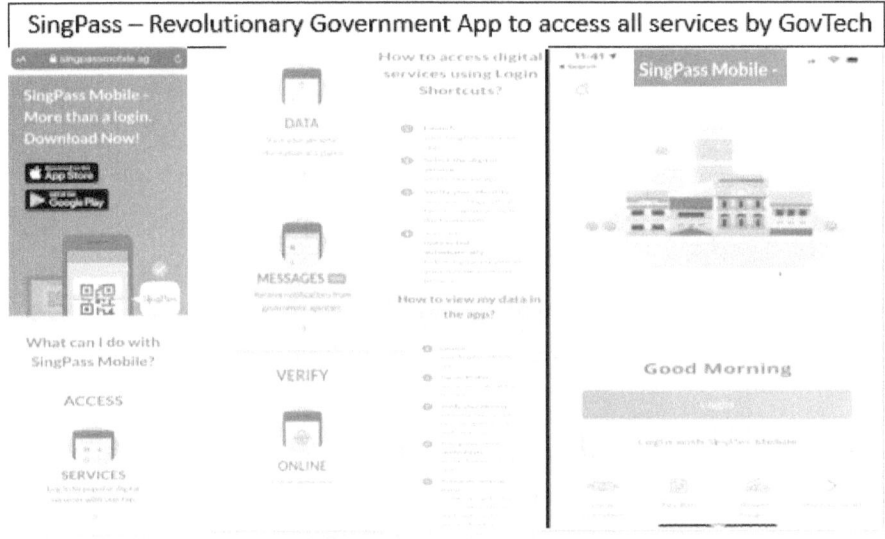

SingPass – Revolutionary Government App to access all services by GovTech

Fig 6.3.2 SingPass – One stop access to all Singapore Govt services & Financial transactions

With the National Digital Rail Integrated with Financial Services and FIs aggressively building APIs, Singapore is one of the first countries in the world to have a comprehensive digital stack making life convenient and secure for its citizens, a true progress of Digital Transformation.

CHAPTER 6.4: Planning Infrastructure with 50 year Perspective into the Future

Singapore is surrounded by sea and is having one of the highest densely populated localities.

To accommodate for increasing population, Singapore periodically undertakes Land Reclamation projects that have increased the land area by over 20% in the past 5 decades.

In 20 years Singapore converted Marina Bay into a vibrant area with mixed usage for corporate and recreational usage being used 100% of the time 7 days a week instead of 5 days for office use

Fig 6.4.1 House: Zero Carbon House

It emphasises on Vertical space utilisation through tall buildings and sky rises. Millions of apartments of highest quality have been constructed.

To increase utilisation of this space, idle times in corporate dwellings is minimised through construction of 'mixed' spaces that encourage 24 by 7 activity through the year by a combination of commercial, official and recreational facilities in the same hub.

One of the key persons behind this visionary architecture is Cheong Koon Hean of HDB.

Cheong Koon Hean is Key architect of Singapore's cityscape and housing estates.

As the head of Singapore's statutory body for Housing development, HDB, the trained architect, Ms. Cheong Koon Hean, is instrumental behind many of the prestigious projects of Singapore like Marina Bay precinct, a signature skyline with a vibrant live-work-play destination, Jurong Lake District, and also the Architecture and Urban Design Excellence Programme.

City in transition from almost nothing in 50 years with millions of high rise aflats and fantastic corporate spaces

Fig 6.4.2 Transitions from almost nothing in 50 years with millions of high rise

HDB built over 1 million apartments with a balance of Very tall vertical high rises and large spaces offering huge green spaces, recreational facilities and tourism attractions for high quality of life for citizens, while, turning Singapore into a magnet for best talent and tourists across the world.

No Traffic congestion despite highest population density through multimodal transportation...

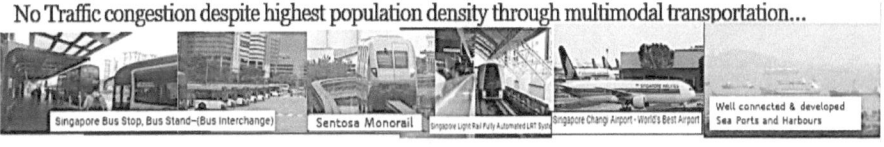

Fig 6.4.3 No Traffic congestion despite highest population density through multimodal transportation

Singapore Government consciously encouraged healthy investment activities of its citizens.

Instead of expensive gas guzzling four wheelers, citizens are encouraged to invest in their own permanent assets like housing. Affordable Public Housing projects by the Government have led to over 80% of its citizens owning their houses while most do not have their personal vehicles due to excessive cost of owning these pocket drainers.

Even the richest in Singapore cannot afford to own and maintain a personal automotive vehicle due to the highest taxes in the world, a conscious decision by the government to channelize investments into productive activities while minimising the stain on the road infrastructure that can lead to a lot of unhappiness & road rage.

Instead, most of the citizens depend on public transportation and shared taxis. The result, tremendous government road & rail infrastructure on ground, above, ground and underground with well connected & maintained, bus & metro, monorail stations. It is a joy to use the public transport in Singapore as, the traffic congestions are kept very low due to optimal motor vehicle traffic.

State of the art underground infrastructure has been planned & constructed to sustain for the next 50 years.

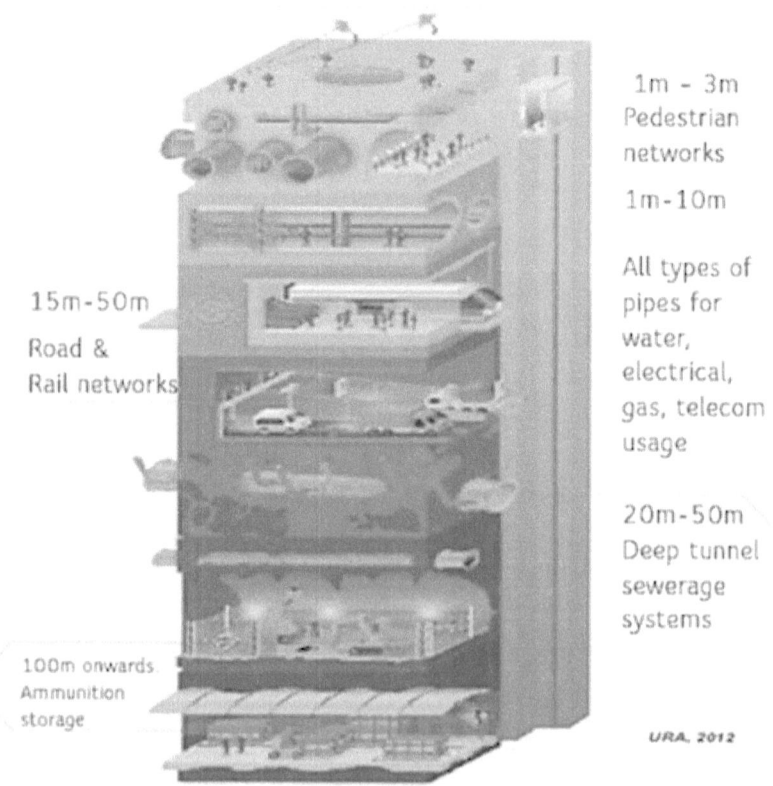

Fig 6.4.4 Singapore Underground Space-Vertical Planning

On-ground transportation will include state of the art isolated lanes for Self-driven cars while protecting the space for cyclists & pedestrians.

Water ways, Drinking water pipes, Drainage pipes, Electric cables, Telecom super ways, Underground metro, highways, subways, Walkways, commercial spaces., warehousing & storage, ammunition storage, long term parking of limited use vehicles etc, are constructed underground, an amazing engineering feat.

Fig 6.4.5 Salient features of Singapore's Infrastructure – An Architectural & Engineering marvel

Over ground infrastructure is leveraged for flyovers, railways and in the future, companies like Boeing are planning for urban helicopter & cargo delivery service where large drones would be used for personal as well as cargo transportation. This calls for constructing helipads & drone pads as well as receiving stations on the top of apartments. Every high-rise apartment with a population of over 1000 persons will be equipped with a Drone/Helicopter parking stations to facilitate cargo and human transportation.

CHAPTER 6.5: Digital Government

Singapore Government's digital transformation road map unveils digital and smart Government with an Intelligence driven connected ecosystem with active collaboration between Government, Businesses and Citizens. The activities of these interconnected and collaborating partners in the digital journey are depicted in the following chart.

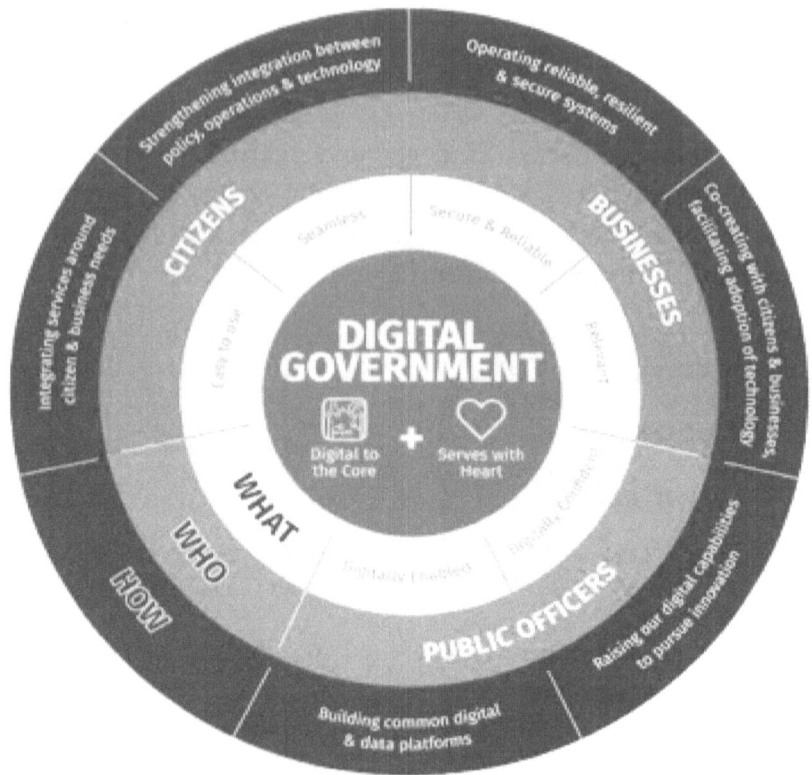

Fig 6.5.1 Singapore Digital Government road map – Roles & responsibilities

Digital Government – Digital to the Core, Serves with Heart

Singapore Government's visionary Digital Government's Blueprint offers a holistic and integrated approach that is strategically coordinated from the office of the Prime Minister. The various dimensions of the comprehensive strategy document are summarised in the following infographic.

Singapore Government's – Digital Government Blueprint – Vision of a Leader

Singapore's visionary, comprehensive and integrated Digital Governance road map designed with an idea to leverage technology to the maximum for the benefit of its citizens while mitigating the negative effects of urbanisation, puts its far ahead of any other Digital Nation groups in the world

2 PRINCIPLES
What guides our actions?

Digital to the Core
Digital Government that uses data, connectivity & computing decisively to re-engineer across processes, re-architect technology operations and transform services for citizens, businesses and public officers.

Serves with Heart
A Digital Government that automates processes where possible so we can better serve citizens with a personal touch in a way that enriches the experience.

3 STAKEHOLDERS
(CITIZENS AND BUSINESSES PUBLIC OFFICERS)

Easy to use
Our digital services are designed to be intuitive, easy to use and accessible anytime, anywhere, on any device.

Seamless
Transactions on our digital services can be completed in a paperless, presence-less manner from start to finish, with the need to provide information only once.

Secure and Reliable
Citizens and businesses are confident their data is secure and our digital services are reliable.

Relevant
Our digital services are designed and built around the needs of citizens and businesses.

Digitally Enabled workplaces
A work environment where they have access to data and digital technologies to design better programmes, collaborate with other public officers to deliver better services and allow us to deploy high quality internal corporate services to be more productive.

Digitally Confident workforce
A workforce with basic digital literacy, and trained to harness data and digital technologies in their work.

6 STRATEGIES
How will we build a digital government?

1. Integrating services around citizen and business needs
We will take a user-centric approach through service journey mapping to design, develop and integrate services around the needs of citizens and businesses.

2. Strengthening integration between policy, operations and technology
We will integrate our policy, operation and technology communities in re-engineering our processes and apply digital technologies (data science, AI, IoT) to transform public services.

3. Building common digital and data platforms
We will develop common, interoperable and easy to use platforms to reduce the time and effort to introduce new digital services. We will set data standards and develop a data architecture to ensure usability of data across Government digital platforms and services.

4. Operating reliable, resilient and secure systems
We will design, build and operate systems against cyber threats and safeguard citizen, business and Government data.

5. Raising our digital capabilities to pursue innovation
We will train public officers to have basic competency in digital skills, proactively manage and deploy ICT talent within the public service, and deepen our technical capabilities through the Centre of Excellence for ICT and Smart Systems.

6. Co-creating with citizens and businesses, and facilitating adoption of technology
We will engage our citizens and businesses to understand their needs, co-create solutions with them, and collaborate with industry to develop new services that are well adopted.

Fig 6.5.2 Two Principles and six Strategies

Digital Government's Key Performance Indicators

KEY PERFORMANCE INDICATORS by 2023 ACROSS VARIOUS INITIATIVES OF THE DIGITAL GOVERNMENT PROJECTS

Fig 6.5.3 Key Performance Indicator by 2023

AI @ Singapore

Artificial intelligence and Machine learning are a major area of interest for the Singapore Government. A national Artificial Intelligence program AISG, is launched with the help of NUS.

AI Singapore (AISG) is a national AI programme launched is by the National Research Foundation (NRF) to anchor deep national capabilities in Artificial Intelligence (AI) AISG is driven by a government-wide partnership comprising NRF, the Smart Nation and Digital Government Office (SNDGO), the Economic Development Board (EDB), the Infocomm Media Development Authority (IMDA), SGInnovate, and the Integrated Health Information Systems (IHiS). (https://www.aisingapore.org/)

The following infographic launched by Singapore Government showcases the various projects launched in line with the National AI program.

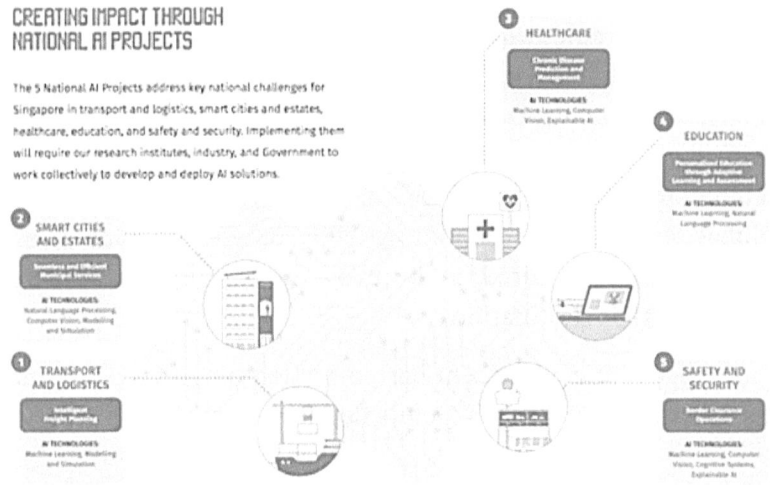

CREATING IMPACT THROUGH
NATIONAL AI PROJECTS

The 5 National AI Projects address key national challenges for Singapore in transport and logistics, smart cities and estates, healthcare, education, and safety and security. Implementing them will require our research institutes, industry, and Government to work collectively to develop and deploy AI solutions.

(Source: Singapore Smart Nation and Digital Government Office)

Fig 6.5.4 Creative Impact Through National AI Projects

In the following sections we will see how Singapore is leveraging AI & ML powered platforms to empower its citizens through innovative solutions in Health care, Smart city transformation and many more.

Blockchain @ Singapore

Blockchain has been recognised as one of the key drivers of Digital transformation journey by Singapore, namely API (Open sourced platforms & Open APIs), Blockchain (Distributed Ledger Technology applications), C (Cloud for scalable & secured solutions across the globe) & Data (AI/ML & Analytics for intelligence embedded applications) & Cyber security for a safe digitalisation & transformation journey.

The Monetary Authority of Singapore (MAS) has been behind many of the innovative Technology applications for the Financial domain. Fraud detection and Prevention, preventing and securing against malicious cyber criminals and to accurately assess the needs of the citizens to device appropriate financial products & solutions is an area of much interest at MAS.

Blockchain with its ability to offer Cyber Risk Management, crash the resource requirement to conduct transactions especially across the border and provide high quality data for analysis is a technology that is of immense value and this has been well recognised by the Singapore Government. Since 2016, Singapore Government and MAS have conducted many experiments, POCs and Pilot projects in collaboration with global leading organisations to substantiate the many benefits of this technology required to leapfrog Singapore into a preferred Financial and Trade destination.

MAS, Singapore government has embarked on a five-phase plan to leverage Blockchain leading to its full-fledged adoption.

Phase 1:
Digitalising
the SGD

Phase 2:
Domestic
interbank
transfer

Phase 3:
Delivery versus
Payment on DLT

Phase 4:
Payment versus
Payment for
cross-border
settlement

Phase 5:
Cross border DvP
and Payment
without Payment

Fig 6.5.5 Phase plan by Monetary Authority of Singapore for Blockchain

MAS is taking an active role in experimenting with disruptive technologies like Blockchain, where it is adopting a five phased approach of moving the killer applications like Cross Border Trade and currency exchange from idea stage to production stage. MAS began experimenting with Blockchain by launching Project UBIN, to digitise Singapore Dollar and use a DLT ledger to transfer these Digital dollars across the banks in the country. After successful experimentation it has now cleared multiple stages of evolution to now be ready to adopt this technology for a number of use cases like Cross border remittances, connecting Singapore exchange with global trade and financial ecosystems, trade network exchanges for leveraging Distributed ledger technology for instant clearing and settlement.

PROJECT UBIN PHASE (1 – 5) by Monetary Authority of Singapore's Project for evaluating applications of Blockchain for clearing and settlement in Cross Border Trade, across Currencies

Courtesy: The Monetary Authority of Singapore (MAS) and Temasek report "Project Ubin Phase 5: Enabling Broad Ecosystem Opportunities".

Project Ubin is a collaborative project with the industry to explore the use of blockchain and distributed ledger technology for the clearing and settlement of payments and securities with a goal to develop simpler-to-use and more efficient alternatives to today's systems, and that are based on central bank-issued digital tokens.

Phase 1 MAS, R3 & consortium of financial institutions on a PoC to conduct inter-bank payments using blockchain technology.

Phase 2 MAS and the Association of Banks in Singapore (ABS) led the successful development of software prototypes of three different models for decentralised inter-bank payments and settlements with liquidity savings mechanisms..

Phase 3: Delivery versus Payment (DvP) MAS and Singapore Exchange (SGX) collaborated to develop Delivery versus Payment (DvP) capabilities for settlement of tokenised assets across different blockchain platforms, and defined a market framework that governs post-trade settlement processes such as arbitration.

Phase 4: Cross-border Payment versus Payment (PvP) The Bank of Canada (BoC), linked up their respective experimental domestic payment networks, namely Project Ubin and Project Jasper, and conducted a successful experiment on cross-border and cross-currency payments using central bank digital currencies.

Phase 5: Enabling Broad Ecosystem Opportunities , the final phase developed the multi-currency payments model described in Phase 4, conducted connectivity testing with other blockchain applications & proved the business value of a blockchain-based payments network.

Fig 6.5.5 (2) 5 stages of Project UBIN by Monetary Authorities of Singapore

Promoting Innovation Through Adoption of Regulatory Sandbox

Singapore Government and MAS (Monetary authority of Singapore) take an active role in encouraging innovation and experimentation through 'Regulatory Sandbox.' All selected ideas are put through an experimentation through active investment and participation by the MAS to see the viability of the new products, services and platforms & also to ensure that there is no risk in adopting the ideas. The models are tweaked to ensure that the risk is minimised, investments are optimised & the output is maximised. Once the idea is successfully cleared by the Sandbox framework, full effort is put behind it to scale up and make it a production grade application. This has resulted in Singapore becoming a hot bed of innovation and made it a magnet for best of entrepreneurs across the globe with the Government acting as active facilitator & supporter. Cross-border settlement of payments & securities with Singapore's currency over Blockchain (DvPvP) is a key milestone being envisioned by MAS.

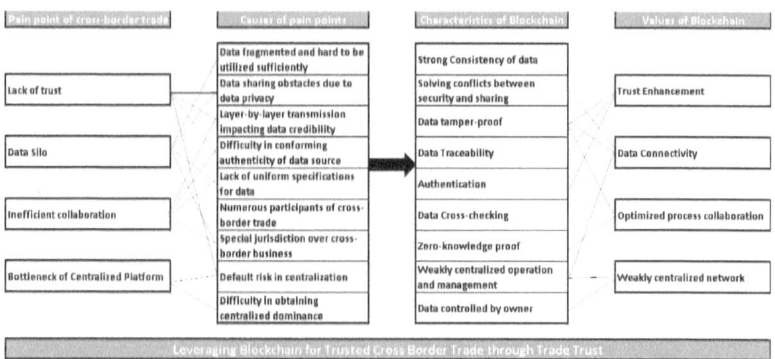

Fig 6.5.6 Leveraging Blockchain for Trusted Cross Border Trade through Trade Trust

GTCN (Global Trade Connect Network)

Through its platform GTCN (Global Trade connectivity network) that connects all the actors in a supply chain & international trade ecosystem, Singapore is envisaging a drastic reduction in complexity, time taken, cost of transactions for exchanging goods & services across the borders. Through DLT platform, Singapore plans to offer a system of settling International assets in other countries with its Digital Singapore dollars for instantaneous, safe, and secure transactions.

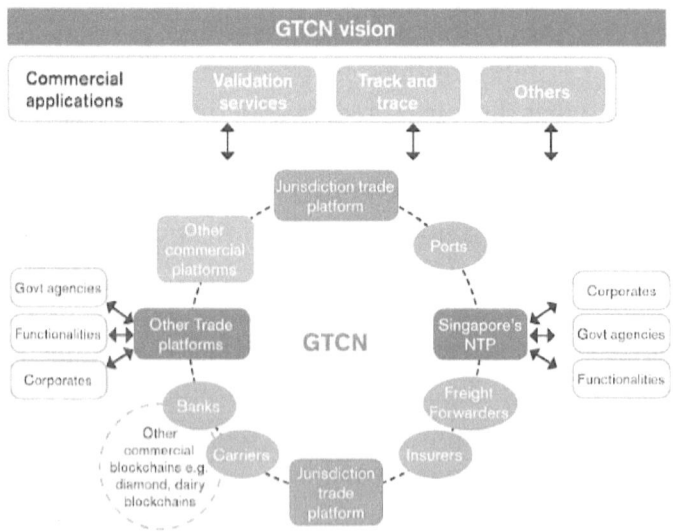

Fig 6.5.7 GTCN Vision

The Monetary Authority of Singapore (MAS) signed a Memorandum of Understanding with its Hong Kong counterpart in November 2017 to jointly develop a cross-border distributed ledger technology (DLT) based infrastructure that will link up digital trade platforms and the growing number of trade-related DLT platforms and communities around the world.

CHAPTER 6.6: Smart Nation Initiatives for all Round Development

Singapore is ranked the world's smartest city as per IMD Smart City Index, an index that measures the way the technology is used to improve the quality of people's lives while reducing the shortcoming of urbanization.

Singapore's activities under the Smart Nation Program are categorized under the following dimensions:

Fig 6.6.1 Smart Nation's Many Initiatives make Singapore the World's Smart City

The Smart Nation and Digital Government Office (SNDGO), under the Prime Minister's Office (PMO), plans and prioritizes key Smart Nation projects, drives the digital transformation of government, builds long-term capabilities for the public sector, and promotes adoption and participation from the public and industry, to take a collective approach in building a Smart Nation. Together with the Government Technology Agency (GovTech), the implementing agency of SNDGO, the group is known as the Smart Nation and Digital Government Group (SNDGG). SNDGG is led by Permanent Secretary Mr. Ng Chee Khern and Deputy Secretary Mr. Tan Kok Yam.(Ref. https://www.smartnation.gov.sg/)

Public Digital Platform to Power the Connected City

Singapore is building a public digital platform and an operating system to which all the public agencies are connected through a secure, sensor network providing anonymised data collected across various facets of their services to citizens.

The data is collected and analysed by leveraging bigdata, cloud, AI & ML technologies to enable the Government to provide better services to its citizens, The analysed information is also shared with the public and other public and private entities for improved decision making & for better planning across the dimensions of transportation, healthcare, education, work etc.

The various activities coordinated under the different dimensions mentioned above, through the SMART Nation initiatives are outlined in the following infographic.

Projects Under SMART Nation Initiative – A Snapshot

Fig 6.6.2 Snapshot of SMART Nation's activities across different dimensions

CHAPTER 6.7: Educating the Smart Nation

Education system in Singapore is designed to develop the students in an all-round manner. All the citizens are required to go through a compulsory military stint that not only develops their personality & inculcates patriotism, but also enables them to grow into responsible and mature human beings. Singapore boasts of over 97% English literacy. Technology is leveraged extensively in education to make it very effective and the learning more fun.

Extensive Use of ICT in Education from Early Stage to Late in Life

The Ministry of Education (MOE) in Singapore encourages the use of Information & Communication Technology in school education right from the age group of 4. The guidelines offered to the schools for the same are outlines as follows:

1. The use of ICT should complement children's learning experiences and be developmentally appropriate.

2. The use of ICT should be facilitated and guided by teachers.

3. The use of ICT should be carefully considered to ensure the safety and well-being of children.

Source: https://www.nel.sg/qql/slot/u143/Resources/Downloadable/pdf/(MOE)ICT%20Guidelines_Final.pdf

Fig 6.7.1 Using ICT enhanced education for educating children from age 4+

Ministry of Education provides extensive support to the schools in adopting to the new age technologies and training the digital native children who are born in the age of smart phones and multimedia.

It provides support to the school in the following ways:

1. Training the teachers and school managements in understanding new technologies & to learn how to leverage them to deliver effective blended learning to the students.

2. Integrating Multimedia into the educational curriculum to develop digitally enhanced education methodologies and create an optimal blended learning.

3. Educating about the moral and ethical aspects of using ICT to ensure responsible applications and

4. Encourage the schools to innovate and evolve better practices to impart effective education using ICT.

Technology-enabled (Tech) toys

Fig 6.7.2 Technology – enabled (Tech) toys

The children are taught, from their very first year in school, through technology enabled toys. Playing using technology to learn concepts like sequencing and matching the right sequences to get the desired outcomes makes learning fun and interesting at the same time very sticky too.

As children get comfortable with Technology from early life, they will develop into Highly competent professionals and entrepreneurs very much needed to keep Singapore ahead in the 21st century

Globally Renowned Higher Educational Institutions in Singapore

For a small country of around 6 million residents, Singapore boasts of an amazing quality of higher educational institutions consistently ranked in the top 100 & sought after by students across the world.

Excellent Educational, Research Institutions- Magnet for the finest students across the world

Fig 6.7.3 Excellent Educational, Research Institution

Public universities offering different disciplines and degrees, financed by the government are open to both Singaporean and international students. The following is a list of Singapore's globally renowned institutions offering top notch educational facilities and learning & research environment:

1. The National University of Singapore (NUS)

2. Nanyang Technological University (NTU)

3. Singapore Management University (SMU)

4. Singapore Institute of Technology (SIT)

5. Singapore University of Technology and Design (SUTD)

6. INSEAD Singapore (Private)

The Universities act as the testbed for advanced technology applications with incubation centres & industry/academia collaborations. Singapore's educational system has adopted a Blockchain platform to offer authentic credential certificates to the students.

OpenCert for Issuing Blockchain Based Certificates to Singapore Students

The platform Open Cert depicted below is a great example of Singapore's adoption of emerging technologies like Blockchain.

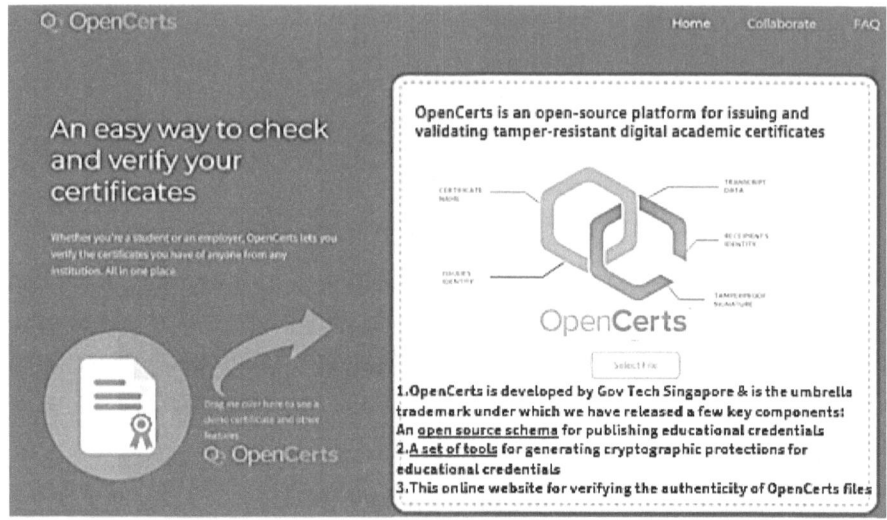

Fig 6.7.4 OpenCerts

OpenCerts platform is used by almost all leading educational institutions in Singapore. The details of Singapore's educational institutions registered for issuing Blockchain based certification is available at https://opencerts.io/registry. This ensures that the entire student eco-system is now familiar with this technology that is considered a hype or being esoteric in most other parts of the world. OpenCerts code base is opensourced and is available at the link https://github.com/OpenCerts.

Collaboration Between Industry & Academia for Disruptive Solutions

Singapore Government nurtures a lot of start-ups and enables a collaboration between public universities and bleeding edge companies working on solutions for leap frogging into digital leadership and benefitting the citizens. Blockchain Living Lab is one such example of Industry-Academic collaboration involving Smart Mesh Foundation.

Fig 6.7.5 Blockchain Living Lab

Project SUNSHINE has the goal of helping to address the SDGs. To be effective, SUNSHINE must be sustainable itself, which is achieved due to the following technology and business models:

- Space-Ground Integration Network

- HyperMesh Infrastructure for the Underserved

- Working for the UN SDGs Generates Wealth in the Community

- Keeping the Wealth in the Community

- Incentivizing Operations through Shared-ROI

SmartMesh Foundation, MeshBox Foundation, and ecosystem partners implementing HyperMesh Protocol Stack (https://smartmesh.io/Space-GroundIntegrationNetworkforSUNSHINE.pdf)

SmartMesh has added a 4th dimension, with the HyperMesh blockchain, with layering of Mesh communication, renewable energy infrastructure, blockchain based applications, and the ability to operate with intermittent or non-existent connection to the Internet. SmartMesh Foundation has implemented Token switching with the world's first Green (low energy) blockchain mining protocol, the Spectrum public blockchain. Spectrum provides a solid Fintech Blockchain foundation on top of which Smart-Contracts, Middleware, and Applications, such as the Tango Distributed/Decentralized Application (DApp) are constructed.

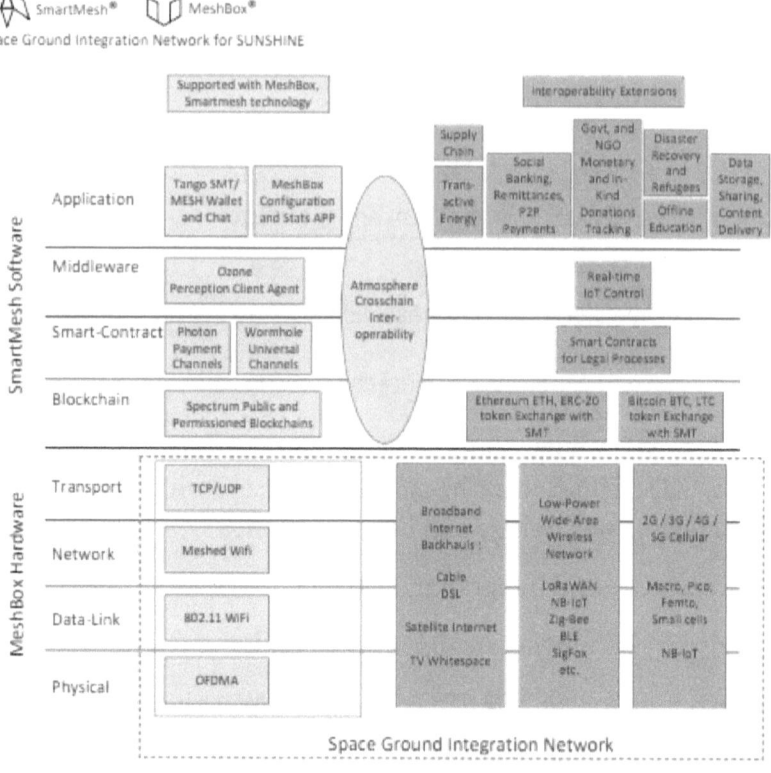

Fig 6.7.6 Architectural components of SGIN and their interconnections

SmartMesh has also developed the world's first off-chain capable Payment Network, Photon for P2P payments, which offloads most transactions on Spectrum. Photon is a Smart Contract running on Spectrum and is the first Blockchain TPS scalability architecture to work even when not connected to its blockchain, Spectrum. (**Source: Smartmesh.io**)

CHAPTER 6.8: Towards Food Sustainability – Innovations Galore

A high density of population and limited land and other natural resources lead to a huge shortage of food production for Singaporeans. However, as they say, 'Necessity is the mother of invention.' Singapore is leveraging technology enabled intelligent farming methods and new generation systems like Hydroponic farming to race towards self-sufficiency in food production in a limited land with highest density. The Government is leaving no stone unturned to follow practices that multiply farming land (through vertical scaling) & Productivity of farming investments!

Hydroponics & Vertical Farming without Sun Light

Sustenair- Hydroponics & Vertical farming without sun light

Fig 6.8.1 Hydroponics & Vertical farming without sun light

Controlling air temperature & humidity, light wavelength, oxygen content in water, CO_2 saturation to enable targeted vegetable & fruit farming with high level of efficiency, production while minimising resource requirements.

Advanced Technology to Intelligently Design Food with Right Nutrition

Fig 6.8.2 Leveraging IOT, AI, ML & Blockchain for intelligent nutrition-based food for patients & elders

Using wearable technology devices, connected to the cloud, Food innovation & Research centre (FIRC) a Government of Singapore undertaking, designs Personalised nutrition for the target group of citizens with different lifestyles & health conditions to keep them healthy & fit. Data from wearables is then fed into ML algorithms, used to design the menu with the right calories and then print them using 3D print technology. Algorithm also predicts potential health issues and suggests appropriate mitigating actions. Leveraging Blockchain Technology, most of the food products available in Singapore departmental stores, offer complete traceability across supply chain, by scan of a QR code pasted on the product's packaging. This ensures that food consumed is always safe from contamination, fakes and crossed expiry dates. Thus, Singapore leverages every opportunity to leverage technology for the wellbeing of its citizens and to attain food sustainability.

CHAPTER 6.9: Enhancing Healthcare and Quality of Life by Leveraging Technology

Singapore being a global trade hub, derives huge amount of revenues from cross border transactions that have been traditionally plagued with high cost and time requirements. Adoption of Blockchain will leapfrog Singapore, not only in improving efficiencies, but also increase its attractiveness in the 'Ease of doing business' index.

Singapore uses advanced technologies in every area where it can impact citizens in a beneficial manner.

For example, Singapore has successfully piloted a project for leveraging Blockchain for settling insurance claims for Pregnant women expeditiously in a fool proof and transparent manner.

Health Tech for Advanced Patient Care, Treatment, and Prevention

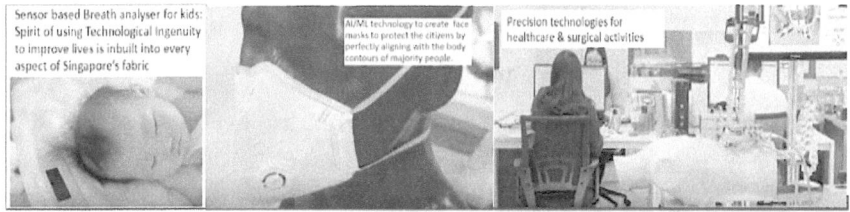

Fig 6.9.1 Leveraging advanced technologies for benefit & wellbeing of patients, elders & the just born

The Healthcare department uses IOT & AI technology to help monitor the breathing of just born babies to ensure that any discrepancies in the breathing patterns that could lead to a high mortality rate can be highlighted and steps taken on the spot. Elder care, Patient monitoring and many areas that affect the quality of life for patients and their wellbeing are always the priority areas for the Government to evolve high technology-based solutions.

nCovid19 – Showing How to Handle Pandemic with World's Lowest Death Rate

In December 2019, a deadly & contagious virus belonging to the family of Corona Viruses spread across China and travelled across the world through Chinese immigrants to countries like Italy and travellers to countries like Singapore in the far east.

Italy became one of the gateways of to Europe and later to the other countries as a large immigrant population from Wuhan reached Italy after their holidays.

The infected Chinese immigrants in Italy socialised extensively with the friendly and warm Italian population, hugging them and kissing them on the streets. Italians being fun-loving and extensive tourists globally, carried the virus within a few weeks across the globe into Asia, Rest of Europe, and America by the end of February.

As the infected passengers travelled across the world, they left multiple traces of the virus on the door handles, toilet seats, fingerprint scanners, restaurants, and restrooms, and inside the flights on the seats and the magazine holders.

The viruses spread across the world rapidly, leaving a death trail close to a million. WHO was blamed for inefficiency in information dissemination with its funding slashed. Leaders of the USA and European countries suffered from the after-effects of the viral attack that was declared as a pandemic. Some countries where the leaders handled the pandemic outcomes in a calibrated manner resulting in lower infections and deaths, rewarded their leaders, and in some other cases, the leaders suffered from negative publicity as the pandemic's impact was seen as a reflection of their maturity, foresight and capability.

Early recognition, blocking entry points, preventing citizen congregation, education & following of personal, hand & cough hygiene, aggressive rapid & free testing, incisive contact tracing, & quarantining of primary & secondary contacts of actual & potential patients, focus on immunity through traditional medicines and supplements for Zinc, Vitamin C & D etc., have enabled countries like Australia, New Zealand, South Korea, Israel, Vietnam, Singapore, Taiwan to successively block & restrict the spread of the epidemic.

Singapore was one of the countries which, despite being at immense risk as a travelling hub and a tourist destination, managed to control the spread of the Pandemic and took care of the citizens in the most laudable manner.

Singapore is leaving no stone unturned, said WHO director-general Tedros Adhanom Ghebreyesus.

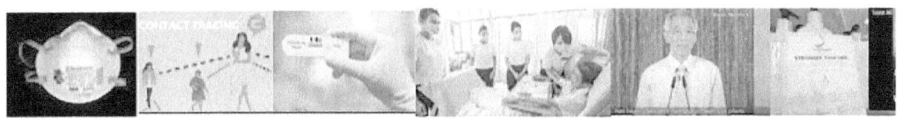

Fig 6.9.2 Singapore adopted the best of practices to limit negative impact of nCovid virus

Being very close to China geographically & ethnically, Singapore is very much exposed to the deadliest contagion with its first nCovid19 case reported on January 23.

Right from day 1, Singapore government put in place a pro-active approach to ensure that the situation never went out of control through aggressive detection, isolation, and treatment of the patients. At the peak of the infection incidence, the non-residents in Singapore, felt it to be safe and secure to stay there, than to leave for their own respective countries.

Pro-active economic measures ensured that the businesses and the citizens never starved for money to survive, by offering direct money transfers to the vulnerable.

Protective face masks & sanitisers are freely & regularly distributed to all the citizens to prevent hoarding of these much-needed items during the time of the crisis.

When the situation seemed to become alarming with discovery of large segments of infected migrant workers, aggressive measures were taken to isolate the patients & provide appropriate treatment, Ministers from the Government closely supervised the treatment and facilities provided to the patients to ensure no slip up of any sort.

The decisive actions undertaken by the Government ensured that the situation is quickly brought under control and with the citizens well educated on personal hygiene and social distancing measures required to keep the virus at bay, the Government slowly relaxed the lock-down measures,

Technology is used to the hilt with the launch of Trace together app that helped in precautionary & preventive activities that limit the spread.

WHO impressed by how Singapore handles coronavirus

The result, Singapore has not only brought the virus under control, but also ensured that the death rate is kept to almost nil at less than 0.1% against the global average of 5%.

Fig 6.9.3 A Temple in Singapore denoting the vibrant cultural heritage

CHAPTER 6.10: Drinking Water Sustenance Through a Multipronged Approach

Singapore has no natural lakes, but reservoirs and water catchment areas have been constructed to store fresh water for Singapore's water supply.

Singapore requires over 2 Billion gallons of drinking water per day for its citizens that is mostly collected from rains over a 400 sqm of catchment area. The high density of population results in a severe shortage and hence there is currently a need for importing most of its drinking water requirements. To overcome this shortage and become self-sufficient in the next 5 years, Singapore is undertaking massive rainwater harvesting, water conservation and desalination projects, effluent treatment & reuse of waste or used water.

Singapore has created a network of criss-crossed water ways and efficient underground drainage systems to collect the rainwater round the year and channelize them into 17 catchment areas across the city. This water is then treated and used for satisfying the demand of the citizens.

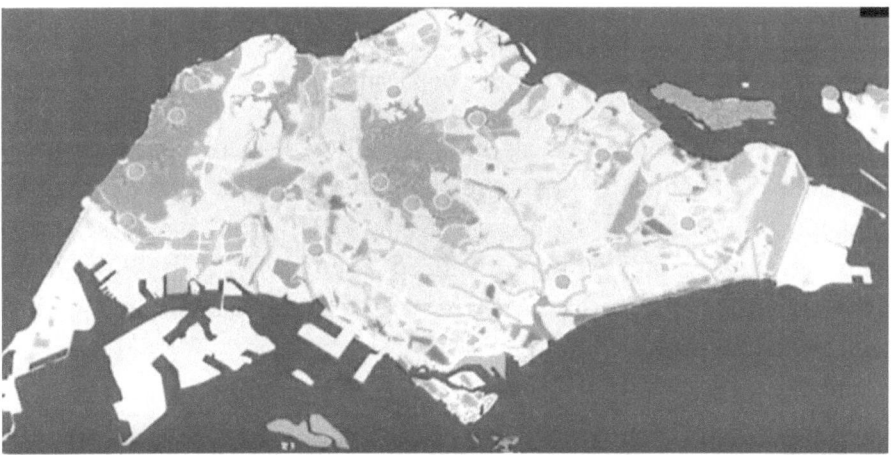

Fig 6.10.1 Criss-crossed Waterways & underground drainage systems collect rainwater into 17 catchment areas across the city

With the country surrounded by Sea all around, it is indeed an irony that Singapore is deprived of a natural resource like drinking water.

To overcome this and to convert the sea water into potable water, Singapore has been exploring Desalination technologies for the past 15 years and has made significant progress that offers a lot of hope for the water starved nations across the world. As they say, in future, the wars are going to be fought to control the access to the rivers offering drinking water.

SINGAPORE'S DESALINATION JOURNEY

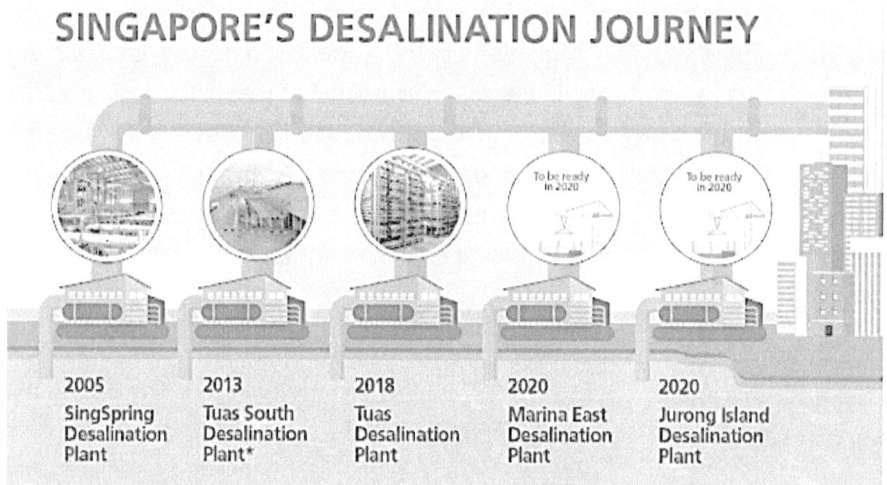

2005	2013	2018	2020	2020
SingSpring Desalination Plant	Tuas South Desalination Plant*	Tuas Desalination Plant	Marina East Desalination Plant	Jurong Island Desalination Plant

Fig 6.10.2 Over 30% of Singapore's drinking water is met through Desalination Plants

Desalination, like NEWater, is a weather-resilient water source, that helps Singapore better cope with the threat of climate change. Singapore Government's National agency, PUB, has turned seawater into drinking water using advanced membrane technology. The government continues to invest in research and technology to find more efficient ways to desalinate seawater.

Its Tuas Desalination plant built at the cost of 217 million Singapore Dollars, produces 30 million gallons per day state of the art fully automated desalination plant, built in 20000 sqm. in 2 years using reverse osmosis methodology has been named the desalination plant of the year 2019.

Fig 6.10.3 Singapore's Tuas 217 million $ Tuas Desalination plant

Since the traditional reverse osmosis process of desalination is energy intensive with one day's requirement of electricity equalling 365000 household requirements, Singapore is in the process of launching the highly energy efficient electrolysis process of desalination through its upcoming projects.

To make desalination more economical, Singapore is investing in new generation technologies employing Electrolysis process (Decomposing the sodium and chloride elements (ions) in the salt by passing electric current. Electrolysis is a technique that uses a direct electric current to drive an otherwise non-spontaneous chemical reaction. This process is very cost effective and is currently in the pilot stage.

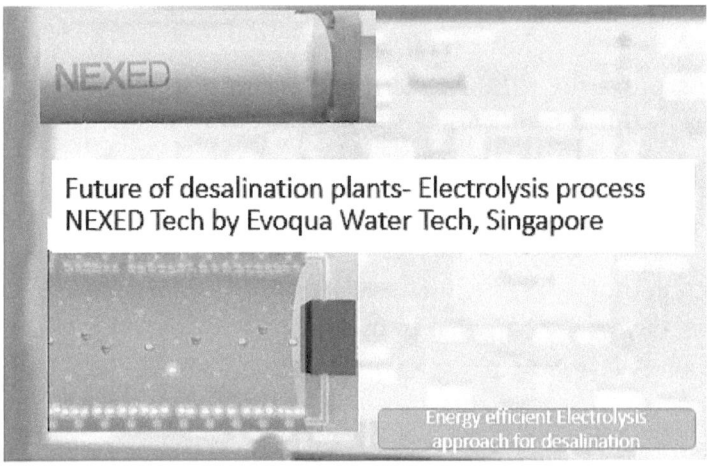

Fig 6.10.4 Electrolysis process for Desalination using NEXED technology

With a combination of actions like optimising water consumption, efficient storing of water in shaded and underground containers preventing evaporation, effluent treatment & recycling of water to the extent possible, rainwater harvesting & desalination, Singapore aims to be self-sufficient in clean water.

CHAPTER 6.11: Dream Tourist Destination for Global Travellers

Travelling by Singapore Airlines to reach Singapore for tourism or business trip is a dream for anyone in the world.

For ages, Singapore presented the best destination and showcased the best, one can experience as a traveller whether it is for business or for recreation or for a symposium or an exhibition.

Singapore leveraged on the locational advantage and built an amazing infrastructure starting from state-of-the-art airports, shopping destinations, sightseeing places and hotels. It is an important junction both by air or sea in case one wants to travel from the west to the Asian countries in the Far east or to Australia or New Zealand.

Changi airport is consistently ranked the best airport in the world. It boasts of an amazing infrastructure housing hotel, parks, destination shopping, recreational hubs, and luxurious sit-outs.

To say that one can take days of joyful recreation to just complete moving around the different terminals is not exaggeration. To reach from one terminal to the other one needs to use a metro.

The important sightseeing destinations that turn Singapore into a Tourist magnet are depicted in the following pictures.

As an author, I was fortunate to spend a lot of time on my first day & the last day of my personal tour to Singapore and enjoy the amazing atmosphere of Jewel Changi airport, which was decked up to greet the Chinese visitors on the occasion of the Chinese New Year being celebrated then.

Fig 6.11.1 Chinese New Year

Singapore is also home to the most automated airport terminal. In Terminal 4, most of the actions and services are undertaken by Robots or in an autonomous fashion without the need of a human intervention.

Fig 6.11.2 Travel and Tourism – The Magnetic charm of Singapore

Tourist attractions

Fig 6.11.3 Tourist attractions

One hopes and envisages that Singapore will continue to evolve as the favourite destination for spending time whether it is for a holiday, sightseeing, business or for staying for long term. This is one of the key factors that Singapore also happens to be a magnet for best talent across the world and draws students & best of Professors to its finest Technology & Business schools.

CHAPTER 6.12: Challenges Faced by Singapore

Like any other country Singapore despite all its progress continues to be bothered by a few challenges.

1. **Dengue outbreaks:** Large proportion of green cover, number of water conservation projects and excessive rainfalls can have a negative repercussion. Due to this, Singapore suffers from high incidence of mosquito borne diseases like Dengue. Government is always gripped with this situation and is now working on leveraging drones to monitor mosquito infested areas and to spray insecticides etc. Strict guidelines are issued to citizens to ensure water does not stagnate anywhere and become a breeding

ground for the mosquitos. Singapore is also trying innovative methods in stopping mosquito reproduction.

Singapore, like China, is trying a second approach, aimed at suppressing the population of mosquitoes by releasing only male mosquitoes. Males infected with Wolbachia are unable to fertilize the eggs of non-infected females.

2. **Increasing cost of living:** Cost of living in Singapore is especially high when it comes to Housing and owning automobiles. While, it is a very planned action by the Government to manage the traffic from becoming chaotic and in reducing the carbon emissions, the high prices are seen to be advantageous for most citizens as they already own houses. Further Government promotes affordable housing projects to most citizens and to migrant workers to ensure that it does not become a burden to them.

3. **Climate change:** Increasing urbanisation and automation could lead to increased carbon emissions and irreversible damage to the environment. An 100% urban area like Singapore city is especially prone to climate warming. To ensure sustainable development in line with UN Guidelines & meet the SDG goals by 2030, Singapore has imbibed a culture of implementing environmentally friendly policies and activities in every sphere of its working. Singapore has committed to ramping up its use of solar energy and greening its land transport sector, with the aim of having nine in 10 peak-period journeys by 2040 made via walking, cycling or rides on public and shared transport. It also wants to make electric vehicles more viable here.

4. **Pollution from nearby countries:** Singapore is surrounded by many industrialised countries and is often subject to attacks of haze and pollution overflowing into its skies from its neighbours. While Singapore offers free face masks to its citizens to ward off the polluted air, this is a problem which is often out of its hands. The commitment of the global organisations to UN Sustainable Development Goals could help in managing the

situation. Singapore is researching a lot on various technologies to curtail pollution and offer the same to the rest of the world that can also help in reducing this stress.

5. **High dependence on international trade and tourists:** Excessive dependence on exports, international trade, and tourism, while being advantageous most of the time, it becomes a great liability during the Pandemic times when the air-traffic cones to a standstill. Prudent economic policies and financial management have enabled the Government in creating a safety net and large reserves to ride out the tough times. However prolonged periods of inactivity could not be sustainable. Hence it is imperative for the Government to device new strategies to get through these unchartered territories and manage the growth for the long term. One idea could be to encourage Singaporean companies and businessmen to ramp up integration with markets across the world, that enables continuous revenue generation even when there is no physical movement.

   Singapore Government has launched Scale-Up SG, a program to groom local businesses to explore and grow their business in international markets. More such steps are needed to ensure that risk of excessive dependence on export – import trade is reduced. Initiatives like the Global ready talent programme are helpful in enabling Singaporeans to get global exposure by spending time in markets like Silicon Valley to get an international perspective and networks.

   Singapore Government also launched the GIA (Global Innovation Alliance) Program to connect local businesses to overseas business and tech communities. GIA Acceleration Programmes will support (i) Singapore Start-ups and SMEs in venturing abroad; and (ii) international start-ups in scaling up in Asia through Singapore as a springboard.

6. **Economic inequalities:** The presence of a flourishing services sector and financial services industry has resulted in attracting highly paid corporate executives. At the same time entrepreneur & start-up friendly policies have resulted in unlocking

substantial value for enterprising promoters and businesses. This has resulted in increasing the income & affordability gap between different segments of population. The successive governments in the country have continuously focused on this to ensure that the lifestyle of the lower income groups is maintained well and in also supporting them during tough times like the nCovid19 outbreak when a large section of the society went without any earnings.

All these problems have been well recognised, and a lot of effort is being undertaken by the successive governments to evolve innovative, determined, and sustainable solutions.

You can easily sense the hardcore belief in every Singaporean, 'We never give up. We are always seeking the next challenge to conquer!'

Conclusion

It is a well-known fact that 'Change is the only constant. Countries and Enterprises who can accept this and adapt themselves to this will be able to survive and thrive in this fast-evolving environment.

Technology is redefining the way Governments engage with their citizens. Government-to-citizen (G2C) services are getting rewired with the help of information and communications technology (ICT).

The advanced automation technologies can blunt the edge of well settled companies & countries and throw up new paradigms and disruptive environments anywhere. Hence, we will see a lot of shifts in the future where large companies that are unable to adapt to change will vanish and new leaders rise in every field. The exponents of Technology can then disrupt any business incumbents who are not vigilant & adaptive to these changes. They will become victims of the 'Frog in the Warm water syndrome'

The new paradigms that have come into place due to nCovid will lead to increased digitisation and lead to more and more businesses moving online.

The organisations will be hard-pressed to sweat their resources more and cut down costs all around. This will entail leveraging advanced Analytics and Automation technologies to predict demand patterns, optimise resources to meet the same etc

IoT, Blockchain, AI/ML, Analytics are combining to transform our lives in unforeseen ways. The emergence, acceptance and now explosion of Cloud based storage and processing has resulted in new paradigms of Digital Transformation across Governments. This has given rise to the concept of Smart Cities that leverage all these technologies for the welfare of the citizens.

We have taken a close look at these emerging technologies like Blockchain, IoT and how they are powering Smart Cities and automation at scale. The movement has just started and there is a long way to go. It is important to understand the various aspects of the Smart Cities and how they are implemented by the best Smart cities and Digital Nations across the world that act as Lighthouses for other nations and cities.

Singapore Government is building a thousand-member team of Digital Ambassador Corps to help micro and small enterprises to embrace digitalisation and to help senior citizens to embrace digital technologies.

Increased digitisation will also lead to availability of humongous data. Planning for the data management, storage, analysis and deriving the insights etc., will be made possible in an accelerated manner.

Cybercriminals will increase their activity to take a share of the increased value being generated and transacted online. Organisations and countries must proactively respond to manage these new threats.

Singapore has proved; what a country that is committed to change and evolves with the times can achieve in this new era.

The 'Heaven' as we all know has added new dimensions and gained technological edge that has allowed it to stay continuously ahead of the rest. It has used all its strengths and achieved a technological edge that makes it the 'Digital Destination.' This also enabled it to take care of its citizens well and proactively ward off challenges posed by new and unforeseen threats like the nCovid19.

To survive and thrive in today's environment, organisations and countries should continuously watch out and be prepared for the disruption. Innovation and Marketing are the key requirements for staying relevant and staying ahead. It is also important to be 'Digital to the Core' and 'Humane at Heart.'

Singapore as a country has set an example for the entire world to follow. It is also well poised to face the challenges that are continuously lurking in the corner.

This is well reflected in the DNA of every Singaporean, who says, 'We will never give up.' No wonder, the country is always looking for 'What Next?'

Lessons from the Digital Transformation Leaders

The COVID-19 pandemic has shaken the world out of its slumber. This pandemic will lead to increased digitisation (i.e. converting information from a physical format into a digital one) and lead to increasing number of businesses moving online. For starters, video conferencing, and virtual meetings for work and educational activities; E-commerce, and distributed ledgers for trusted online transactions will be implemented at rocket speed. There will also be a greater use of robots. From companions, service agents, warehouse workhorses, to drones for surveillance, and delivery agents, a new era beckons; virtual reality in retail, augmented reality in education, and tourism, and finally, the implementation of AI and ML in almost every aspect of our lives, these are a few (of many) advancements that are around the corner.

The organisations will be hard-pressed to sweat their resources more and cut down costs all around. This will entail leveraging advanced Analytics and Automation technologies to predict demand patterns, optimise resources to meet the same etc. Increased digitisation will also lead to availability of humongous data. Planning for the data management, storage, analysis and deriving the insights etc., will be made possible in an accelerated manner.

There is a paradigm shift in the application of technology, with an accelerated digital transformation impact in every sphere of life. The world will now be a lab for experimentation and adoption of all emerging technologies, with the support of regulators. Every advancement also brings its share of challenges, and here, we foresee an increased stress on cybersecurity, with new threats like chemical and biological warfare, and related accidents.

Opportunity for AI in India

According to a recent Accenture study, AI can add US$957 billion (15% of current gross value added) to India's economy by 2035. It's no surprise then, that the government has introduced bold, multi-pronged initiatives to augment labor productivity and innovation with an eye to driving growth. But despite ranking high in terms of the number of AI start-ups, India lags behind other G20 nations, in particular the United States and China, in innovation and tech development in the field. Nevertheless, an allocation of $480 million in Budget 2018 for research, training and skills development in robotics, digital manufacturing, big data intelligence and AI underlines the Indian government's commitment to new technologies that are seen as key to boosting economic and social development. Policy makers have also put together a roadmap for emerging technologies and established a task force, which outlines how AI in India will shape up in the coming future.

Role of Government for Attaining Leadership Role in AI

The following must be the vision for India in the field of AI for the next 10 years. India will be the highest populated country in the world with advantages such as:

1. Large domestic consumption for cutting edge technology services

2. Highest number of English-speaking engineers graduating from universities – these engineers could be oriented in the AI industry.

3. Highest number of connected devices, mobile & cellular connections

4. Largest number of smart cities & integrated command and control centers generating valuable data and consuming the insights provided by AI-ML Systems.

5. Largest numbers of Centers of Excellence across the country's universities collaborating with the global leaders, operating in the cutting edge of technologies including Quantum computing.

6. Highest volume of Data generated by the billions of connected devices, offering a huge value that could be unlocked with the help of disruptive technologies like AI-ML & Blockchain.

7. Backyard to the world for manufacturing hardware for AI industry

8. Trainers to the world for AI adoption across the world

9. Offering cost-effective development ecosystem for AI solutions across the world And Finally,

10. A young population which is tech savvy

The Role of Academic Institutions in Achieving the Targets for Advanced Technologies

India has a very good academic structure when it comes to AI but the outcome in terms of innovation and Intellectual Property (IP) is on a lower side. India is not focusing on creating IPs or Advanced Technology products at the academic level. A center for excellence in this sector will be very helpful but it should be thoroughly monitored by government.

IISc, IIT Mumbai, IIT Kanpur, IIT Patna, IIT Delhi, IIT Madras, some IIITs and a few central universities are leading the efforts of the academia in India. However, the number of research papers and patentable solutions developed in the Indian academic ecosystem is considered miniscule compared to the leading countries in the world like US, China and the D5 nations. The research conducted by the

Indian academic institutions is hampered by the siloed approach and the lack of coordination between the Industry, academia, and Start-up ecosystem & is disjointed from the real-life situations where the difference can be perceptible. The number of solutions developed through the research in India; that has been commercialized on large scale to generate value; is dismal. This also leads to the reluctance of the industry and government to engage with substantial investments that can produce cognizable results. There should be a clear value proposition for the enterprises to partner with the Academia for co-investing in developing disruptive technology solutions.

Role of Blockchain in Digital Transformation

Distributed Ledger technologies like Blockchain enable risk management by offering controlled access, fool proof authentication and trusted authorization for those involved and transactions conducted between them by acting as a trusted third party. Digital transformation at scale can thus be expedited in a secured manner. The following are some of the crucial activities to be done by Government of India to support path breaking applications of Distributed Ledger Technologies:

- Unique Blockchain based self-sovereign digital identities for all citizens

- Central bank digital currency on a national permissioned Blockchain

- Regulatory recognition of data and transaction records stored on Blockchain.

This will help is eliminating corruption & inefficiency linked leakages, thus unlocking huge government resources that may be otherwise wasted or are spent unproductively. Blockchain enables availability of high-quality data for AI and ML applications. Trust and transparency offered by Blockchain, as well as secured private digital identities (of devices and people) offers high quality record keeping, an integral component for several applications in Financial, Supply chain and in Health care applications like clinical record management, and electronic health records administration.

Reinventing Careers Through Re-Skilling

Realising that change is the only constant in today's fast evolving world, knowledge professionals should upskill themselves to stay relevant in a world. Some of the activities that are suggested are outlined below:

1. Keep abreast of the latest use cases by studying the way the global leaders like Amazon, Netflix, Facebook, Google, JP Morgan, HSBC etc., are leveraging emerging technologies.

2. Keep abreast of the tools and platforms offered by leading companies in the field like IBM, Google, Microsoft, Intel etc.

3. Undertake courses on Online platforms to upskill on the trending topics in the emerging technologies.

4. Study how AI/ML & Blockchain applications are being implemented in leading countries like China, USA, Singapore & Middle East.

5. Always look proactively for opportunities to implement AI/ML solutions in practical scenarios while being sensitive to the way, various businesses are leveraging the same.

6. Learn to be computer literate and try to pick up a language like Node JS, Golang or Python.

7. Try to write articles, white papers or books that will force one to conduct intense research around related topics.

Role of Government in Encouraging Skilling, Reskilling and Upskilling

There is a large need to take care of the existing workforce that is likely to be displaced from their jobs due to the emerging technology led automation. Professionals have a dire need to re-skill & stay relevant. Hence it is imperative for the Government to put in place a concerted strategy to deliver appropriate competence & impart the industry relevant skill while encouraging the learners to master the same.

It is imperative to partner Global leading educational/training providers and offer the highest quality programs that combine theoretical, practical, and cutting-edge solutioning capabilities to the executives across the various stages of the corporate lifecycle.

1. Reskilling should be encouraged through proper incentives and opportunity for career growth

2. Formal and informal education with highest standards should be made available with proper standardization and recognized certification that is valued by industry.

3. Online education through MOOCs should be blended with real life opportunities to explore and implement solutions on job.

In conclusion, while it is true that COVID-19 pandemic has resulted in a dramatic paradigm shift in the form of increased digitalisation & automation, times have also become very challenging due to the availability of multiple resources across the software development lifecycle with the proliferation of open source technologies and highly secure and scalable cloud enabled SAAS environment.

While countries like India have understood the paradigm of Smart Cities and Digital Governance, they have a long way to go as far as implementation is concerned, It is here that the examples set by countries like Estonia, Dubai (UAE), Singapore (which is also the Smartest city in the world) have to closely studied, understood and evaluated for implementation.

Sustainable Development is very much required if the world has to survive into the future and we manage to handover a safe and secure world to our younger generations.

Hence we need to consciously strive to achieve Sustainable development like the way Singapore is planning 50 years in advance and is overcoming a number of geographic limitations as well as resource constraints. It is indeed possible to overturn the tide of environmental degradation caused by industrial development and this has been stressed in every chapter in this book through real life examples that are in vogue.

Digital Nations and Smart Cities are rapidly evolving, and the resulting digitalization is leading to several benefits while also exposing the citizens to unforeseen benefits. Blockchain is enabling risk management for a secured automation. The book takes a close look all paradigms of Digital Governance while relating to the application of these principle in real life through the case study of Singapore, which is one of the top 3 densest, but also, is one of the most sustainable cities. This book will be a useful resource for professionals, consultants, government servants and students who wish to come to grip with the emerging technologies & their applications in governance and play an active role in community building activities.

The book explores the emergence, evolution & adoption of advanced digital technologies like IoT, Analytics, Blockchain for improved governance, sustainable development & better quality of life and happiness for citizens across the world.

Srinivas Mahankali is a passionate advocate and practitioner of Blockchain technology. An alumnus of IIT Madras and IIM Bangalore, He worked in leading Indian and Multinational organizations in India at various Top management roles. He led the Digital transformation activities and IT organizations in some of India's leading organizations. Srinivas authored/ co-authored multiple books on emerging technologies. His book Blockchain the Untold Story is the first book in the world to be translated into Chinese language using AI agents.

Price Rs 395/00
ISBN 978 1 97991 140 7

Xpress Publishing
An Imprint of Notion Press

www.ingramcontent.com/pod-product-compliance
Lightning Source LLC
Chambersburg PA
CBHW021401210526
45463CB00001B/181